NOVEL WATER TREATMENT AND SEPARATION METHODS

Simulation of Chemical Processes

NOVEL WATER TREATMENT AND SEPARATION METHODS

Simulation of Chemical Processes

Edited by
Bharat A. Bhanvase, PhD
Rajendra P. Ugwekar, PhD
Raju B. Mankar, PhD

APPLE
ACADEMIC
PRESS

Apple Academic Press Inc.　　　　Apple Academic Press Inc.
3333 Mistwell Crescent　　　　　　9 Spinnaker Way
Oakville, ON L6L 0A2 Canada　　　Waretown, NJ 08758 USA

© 2018 by Apple Academic Press, Inc.

First issued in paperback 2021

No claim to original U.S. Government works

ISBN-13: 978-1-77463-650-3 (pbk)
ISBN-13: 978-1-77188-578-2 (hbk)

Library and Archives Canada Cataloguing in Publication

Novel water treatment and separation methods : simulation of chemical processes / edited by Bharat A. Bhanvase, PhD, Rajendra P. Ugwekar, PhD, Raju B. Mankar.
"This book, Novel Water Treatment and Separation Methods: Simulation of Chemical Processes, is the outcome of national conference REACT-16, organized by the Laxminarayan Institute of Technology, Nagpur, Maharashtra, India, in 2016."--Preface.
Includes bibliographical references and index.
Issued in print and electronic formats.
ISBN 978-1-77188-578-2 (hardcover).--ISBN 978-1-315-22539-5 (PDF)
1. Sewage--Purification. I. Bhanvase, Bharat A., editor II. Ugwekar, Rajendra P., editor III. Mankar, Raju B., editor

| TD745.N68 2017 | 628.3 | C2017-903824-9 | C2017-903825-7 |

Library of Congress Cataloging-in-Publication Data

Names: Bhanvase, Bharat A., editor. | Ugwekar, Rajendra P., editor. | Mankar, Raju B., editor.
Title: Novel water treatment and separation methods : simulation of chemical processes / editors, Bharat A. Bhanvase, PhD, Rajendra P. Ugwekar, PhD, Raju B. Mankar.
Description: Toronto ; Waretown, NJ : Apple Academic Press, 2017. | "Outcome of national conference REACT-16, organized by the Laxminarayan Institute of Technology, Nagpur, Maharashtra, India, in 2016"--Introduction. |
Includes bibliographical references and index.
Identifiers: LCCN 2017025355 (print) | LCCN 2017030043 (ebook) | ISBN 9781315225395 (ebook) | ISBN 9781771885782 (hardcover : alk. paper)
Subjects: LCSH: Sewage--Purification--Congresses. | Water--Purification--Filtration--Congresses. | Extraction (Chemistry)--Congresses. | Water--Pollution--Congresses.
Classification: LCC TD745 (ebook) | LCC TD745 .N68 2017 (print) | DDC 628.1/66--dc23
LC record available at https://lccn.loc.gov/2017025355

Apple Academic Press also publishes its books in a variety of electronic formats. Some content that appears in print may not be available in electronic format. For information about Apple Academic Press products, visit our website at **www.appleacademicpress.com** and the CRC Press website at **www.crcpress.com**

ABOUT THE EDITORS

 **Bharat A. Bhanvase, PhD,** is currently working as Associate Professor in the Chemical Engineering Department at the Laxminarayan Institute of Technology, RTM Nagpur University, Nagpur, Maharashtra, India. His research interests are focused on conventional and cavitation-based synthesis of nanostructured materials, ultrasound-assisted processes, polymer nanocomposites, heat transfer enhancement using nanofluid, process intensification, and microreactors for nanoparticle and chemical synthesis. He has published 53 articles in international and national journals and has presented papers at international and national conferences. He has written 19 book chapters in internationally renowned books and applied for two Indian patents. He is the recipient of a Summer Research Fellowship from the Indian Academy of Sciences, Bangalore, India, in 2009 and Young Scientists (Award) from Science and Engineering Research Board, Department of Science and Technology in the year 2015. He has more than 13 years teaching experience. He has completed a research projects with the University of and currently handling three research project sanctioned by Rashtrasant Tukadoji Maharaj Nagpur University and Science and Engineering Research Board, New Delhi. He is a reviewer for several international journals. Dr. Bhanvase completed his BE in Chemical Engineering from the University of Pune, his ME in Chemical Engineering from Bharati Vidyapeeth University Pune, and his PhD in Chemical Engineering from University of Pune, India.

 **Rajendra P. Ugwekar, PhD,** is currently working as Associate Professor and Head in the Chemical Engineering Department at the Laxminarayan Institute of Technology, RTM Nagpur University, Nagpur, Maharashtra, India. His research interests are focused on hydrogen energy, nanotechnology, wastewater treatment, and heat transfer enhancement using nanofluid. He has published several articles in international and national journals and has presented papers at international and national

conferences. He has more than 22 years of teaching experience and several years of industrial experience as well. He has completed research projects received from AICTE, India of Rs. 24 Lakhs and NSTEDB, New Delhi and UGC and has several advanced students working with him. He has worked as Head of the Department at Anuradha Engineering College, Chikhali, Buldhana, India, and at the Priyasharshani Institute of Engineering and Technology, Nagpur, India. He is a trainer and motivator for entrepreneurship.

Dr. Ugwekar has completed his B.Tech in Chemical Engineering from Nagpur University, his MTech in Chemical Engineering from Nagpur University, Nagpur, and his PhD in Chemical Technology from Sant Gadge Baba Amravati University, Maharashtra, India.

 **Raju B. Mankar, PhD,** is currently Director of the Laxminarayan Institute of Technology in Nagpur, India. He was also the former Vice-Chancellor of Dr. Babasaheb Ambedkar Technological University in Lonere-Raigad, India (2010–15). Professor Mankar was the Nominee of His Excellency, the Vice President of India on the Court of Pondicherry University during (2007–2010). He is Member of Western Regional Committee, Mumbai and also a Member of Admissions Regulating Authority, Government of Maharashtra. He has been nominated as a Council Member by the President of the Institution of Engineers (India), as an "Eminent Engineering Personality". Professor Mankar has served as Vice-Chairman of the Governing Council, Engineering Staff, College of India, Hyderabad (2012-15). Professor Mankar is a Fellow of the Institution of Engineers (India), a Life Member of Indian Institute of Chemical Engineers, and a Life Member of Indian Society for Technical Education. He is also Fellow of DAAD, Federal Republic of Germany.

Dr. Mankar earned his MTech in Chemical Engineering from the Laxminarayan Institute of Technology, Nagpur, and his PhD from the Indian Institute of Technology, Kanpur, India. He is the recipient of several prestigious awards and has worked as a Visiting Scientist at the University of Karlsruhe.

CONTENTS

LIST OF CONTRIBUTORS

S. V. Admane
Department of Chemical Engineering, MIT Academy of Engineering, Alandi (D), Pune, Maharashtra, India

S. S. Anwani
Laxminarayan Institute of Technology, Rashtrasant Tukadoji Maharaj Nagpur University, Nagpur 440033, Maharashtra, India

D. P. Barai
Laxminarayan Institute of Technology, Rashtrasant Tukadoji Maharaj Nagpur University, Nagpur 440033, Maharashtra, India

B. A. Bhanvase
Chemical Engineering Department, Laxminarayan Institute of Technology, Rashtrasant Tukadoji Maharaj Nagpur University, Nagpur 440033, Maharashtra, India. E-mail: bharatbhanvase@gmail.com

R. P. Birmod
Department of Chemical Engineering, Laxminarayan Institute of Technology, Rashtrasant Tukadoji Maharaj Nagpur University, Nagpur 440033, Maharashtra, India

A. B. Bodade
Department of Chemistry, Nanotechnology Research Laboratory, Shri Shivaji Science College, Amravati 444602, Maharashtra, India

O. D. Borole
Department of Chemical Engineering, Sinhgad College of Engineering, Vadgaon Budruk, Pune 411041, Maharashtra, India

G. N. Chaudhari
Department of Chemistry, Nanotechnology Research Laboratory, Shri Shivaji Science College, Amravati 444602, Maharashtra, India

S. P. Chaudhari
Department of Chemistry, Nanotechnology Research Laboratory, Shri Shivaji Science College, Amravati 444602, Maharashtra, India

S. M. Chavan
Department of Chemical Engineering, Sinhgad College of Engineering, Vadgaon, Pune, Maharashtra, India

P. V. Chavan
Department of Chemical Engineering, Bharati Vidyapeeth Deemed University, College of Engineering, Pune 411043, Maharashtra, India. E-mail: pvcuict@gmail.com; pvchavan@bvucoep.edu.in

S. D. Dawande
Department of Chemical Engineering, Laxminarayan Institute of Technology, Nagpur, Maharashtra, India

M. P. Deosarkar
Chemical Engineering Department, Vishwakarma Institute of Technology, Upper Indira Nagar, Bibvewadi, Pune 411037, Maharashtra, India

P. Dhanke
Chemical Engineering Department, PVPIT, Sangli, Budhgaon 416304, Maharashtra, India

S. G. Gaikwad
Chemical Engineering and Process Development Division, National Chemical Laboratory, Pashan, Pune 411008, Maharashtra, India
CEPD Department, National Chemical Laboratory, Pune, Maharashtra, India

V. N. Ganvir
Department of Petrochemical Technology, Laxminarayan Institute of Technology, Nagpur, Maharashtra, India

S. Ghodke
Department of Chemical Engineering, Sinhgad College of Engineering, Pune 41104, Maharashtra, India. E-mail: saghodke@gmail.com

P. R. Gogate
Department of Chemical Engineering, Institute of Chemical Technology, Mumbai 400019, Maharashtra, India. E-mail: pr.gogate@ictmumbai.edu.in

S. Husain
Department of Chemical Engineering, Aligarh Muslim University, Aligarh 202002, Uttar Pradesh, India

Md. N. Iqbal
Department of Food Engineering, National Institute of Food Technology, Entrepreneurship and Management (NIFTEM), HSIIDC Estate, Kundli, Haryana, India

S. Jadhav
Department of Chemical Engineering, MIT Academy of Engineering, Alandi, Pune 412105, Maharashtra, India

P. D. Jolhe
Department of Biotechnology, Sinhgad College of Engineering, Savitribai Phule Pune University, Pune 411008, Maharashtra, India. E-mail: pdjolhe.scoe@sinhgad.edu

A. A. Kadu
Department of Chemical Engineering, Laxminarayan Institute of Technology, Rashtrasant Tukadoji Maharaj Nagpur University, Nagpur 440033, Maharashtra, India

S. P. Kamble
Chemical Engineering and Process Development Division, CSIR-National Chemical Laboratory, Dr. Homi Bhabha Road, Pune 411008, Maharashtra, India. E-mail: sp.kamble@ncl.res.in

R. Kankate
Department of Chemical Engineering, Sinhgad College of Engineering, Pune 41104, Maharashtra, India

N. G. Kanse
Department of Chemical Engineering, FAMT, Ratnagiri, Maharashtra, India

J. Katiyar
Department of Chemical Engineering, Harcourt butler Technological Institute, Kanpur 208002, Uttar Pradesh, India

A. Khandare
Chemical Engineering and Process Development Division, CSIR-National Chemical Laboratory, Dr. Homi Bhabha Road, Pune 411008, Maharashtra, India

P. Kotwal
Department of Chemical Engineering, MIT Academy of Engineering, Alandi, Pune 412105, Maharashtra, India

G. P. Lakhawat
Department of Chemical Engineering, Priyadarshini Institute of Engineering and Technology, Nagpur 440019, Maharashtra, India

M. L. Meshram
Department of Civil Engineering, Laxminarayan Institute of Technology, Nagpur, Maharashtra, India

S. P. Meshram
Department of Nanoscience and Nanotechnology, Centre for Materials for Electronics Technology, Pune 411008, Maharashtra, India

S. P. Nayak
National Institute of Food Technology Entrepreneurship and Management, Plot No. 97, Sector 56, HSIIDC Industrial Estate, Kundli, Sonepat 131028, Haryana, India

V. D. Pakhale
Department of Chemical Engineering, MIT Academy of Engineering, Alandi, Pune 412105, Maharashtra, India. E-mail: vinodpakhale123@gmail.com

R. R. Patil
Department of Petrochemical Technology, Laxminarayan Institute of Technology, Nagpur, Maharashtra, India

N. M. Rane
Department of Chemical Engineering, MIT Academy of Engineering, Alandi (D), Pune, Maharashtra, India

V. Sakhalkar
Chemical Engineering and Process Development Division, CSIR-National Chemical Laboratory, Dr. Homi Bhabha Road, Pune 411008, Maharashtra, India

R. S. Sapkal
UDCT, Sant Gadge Baba Amaravati University, Amaravati, Maharashtra, India

M. D. Sardare
Department of Chemical Engineering, MIT Academy of Engineering, Alandi, Pune 412105, Maharashtra, India

A. Sharma
Department of Food Engineering, National Institute of Food Technology, Entrepreneurship and Management (NIFTEM), HSIIDC Estate, Kundli, Haryana, India

R. Shetty
Chemical Engineering and Process Development Division, CSIR-National Chemical Laboratory, Dr. Homi Bhabha Road, Pune 411008, Maharashtra, India

S. P. Shewale
Department of Chemical Engineering, MIT Academy of Engineering, Alandi (D), Pune, Maharashtra, India

S. D. Shingte
Department of Chemical Engineering, Faculty of Applied Sciences, TU Delft, Delft, The Netherlands

S. R. Shirsath
Department of Chemical Engineering, Sinhgad College of Engineering, Vadgaon Budruk, Pune 411041, Maharashtra, India

S. P. Shirsat
Laxminarayan Institute of Technology, RTM Nagpur University, Nagpur 440033, Maharashtra, India

S. Singha
Department of Food Engineering, National Institute of Food Technology, Entrepreneurship and Management (NIFTEM), HSIIDC Estate, Kundli, Haryana, India
National Institute of Food Technology Entrepreneurship and Management, Plot No. 97, Sector 56, HSIIDC Industrial Estate, Kundli, Sonepat 131028, Haryana, India

S. H. Sonawane
Department of Chemical Engineering, National Institute of Technology Warangal, Warangal 506004, Telangana, India

S. M. Sontakke
Department of Chemical Engineering, Institute of Chemical Technology, Mumbai 400019, Maharashtra, India

M. H. Talwekar
Laxminarayan Institute of Technology, Rashtrasant Tukadoji Maharaj Nagpur University, Nagpur, Maharashtra, India

T. Telang
Department of Chemical Engineering, Vishwakarma Institute of Technology, Upper Indira Nagar, Bibwewadi, Pune 411037, Maharashtra, India

M. A. Thombare
Department of Chemical Engineering, Bharati Vidyapeeth Deemed University, College of Engineering, Pune 411043, Maharashtra, India

A. Thorgule
Laxminarayan Institute of Technology, RTMNU, Amravati Road, Nagpur 440033, Maharashtra, India

R. P. Ugwekar
Laxminarayan Institute of Technology, RTMNU, Amravati Road, Nagpur 440033, Maharashtra, India
Department of Chemical Engineering, Laxminarayan Institute of Technology, Nagpur 440033, Maharashtra, India

V. S. Wadgaonkar
Department of Petrochemical Engineering, Maharashtra Institute of Technology, Pune 411038, Maharashtra, India

S. M. Wagh
Laxminarayan Institute of Technology, Rashtrasant Tukadoji Maharaj Nagpur University, Nagpur, Maharashtra, India

S. Waghela
Department of Chemical Engineering, MIT Academy of Engineering, Alandi, Pune 412105, Maharashtra, India

LIST OF ABBREVIATIONS

AOPs	advanced oxidation processes
AR	acid red
AR	analytical reagent
BET	Brunauer–Emmett–Teller
BI	bio-degradability index
BLM	bulk liquid membrane
CCD	central composite design
CHD	coronary heart disease
D2EHPA	di-2-ethylhexylphosphoric acid
DHA	docosahexaenoic acid
DI	de-ionized
EA	ethyl acetate
EDAX	energy dispersive analysis of X-rays
EDS	energy dispersive spectroscopy
EDX	energy-dispersive X-ray spectroscopy
EFAs	essential fatty acids
ELM	emulsion liquid membrane
EPA	eicosapentaenoic acid
EU	European Union
FDA	Food and Drug Administration
FESEM	field emission scanning electron microscope
FTIR	Fourier transform infrared spectroscopy
HC	hydrodynamic cavitation
HPLC	high performance liquid chromatography
HPVH	high-pressure valve homogenizers
LDH	layered double hydroxides
LLE	liquid–liquid extraction
MAE	microwave-assisted extraction
MB	methylene blue
MBE	molecular beam epitaxy
MgO	magnesium oxide
MOVPE	metal-organic vapor-phase epitaxy
MRI	magnetic resonance imaging
NBE	near band edge emission

NDT	non-destructive testing
NIR	near infrared
NPs	nano particles
PE	petroleum ether
PHE	plate heat exchanger
PL	photoluminescence
PSI	pervaporation separation index
PUFA	polyunsaturated fatty acid
PVA	polyvinyl alcohol
PVC	polyvinyl chloride
RBCs	red blood corpuscles
RF	radio frequency
RhB	rhodamine B
RSM	response surface methodology
RTE	ready-to-eat
SEM	scanning electron microscopy
SLCFB	solid-liquid circulating fluidized bed
SLFBs	Solid-liquid fluidized beds
SLM	supported liquid membrane
SR	swelling ratio
SS	stainless steel
TA	titrable acidity
TGA	thermogravimetric analysis
TIOA	tri-iso-octyl amine
TISAB	total ionic strength adjustment buffer
TOC	total organic carbon
TSS	total soluble solids
UAE	ultrasound-assisted extraction
VORWW	vegetable oil refinery wastewater
WHO	World Health Organization
XRD	X-ray diffraction

PREFACE

This book, *Novel Water Treatment and Separation Methods: Simulation of Chemical Processes,* is the outcome of National Conference REACT-16, organized by the Laxminarayan Institute of Technology, Nagpur, Maharashtra, India, in 2016. We have included the research articles related to applications of chemical processes for selected wastewater treatment examples, separation techniques, and modeling and simulation of chemical processes in a single book that will benefit students, researchers, faculty, and industry professionals.

We have received inspiration from Dr. V. S. Sapkal, Former Vice Chancellor, RTM Nagpur University, Nagpur; the Eminent Scientist Dr. B. D. Kulkarni of National Chemical Laboratory, Pune; Professor A.B. Pandit, Institute of Chemical Technology, Mumbai; Prof. Arun S Moharir, Professor, Chemical Engineering Department, I. I. T. Bombay; and Dr. Swapneshu Baser, Director, Deven Supercriticals Limited, Mumbai. They have inspired us to take up the assignment of editing this book. This book covers different areas of chemical engineering and technology.

The important and recent topics in the field of chemical engineering and technology including modeling and simulation, wastewater treatment, separation techniques, etc., have been presented in this book. We, the editors, are pleased to bring out this special type of book that covers above-mentioned areas.

We would also like to acknowledge the team of Apple Academic Press, Mr. Ashish Kumar, President and Publisher, and Mr. Rakesh Kumar from Apple Academic Press. Inc., USA, for their prompt and supportive attention to all our queries related to editorial assistance.

Dr. R. B. Mankar, Former Vice Chancellor, Dr. Babasaheb Ambedkar Technological University, Lonere and present Director of Laxminarayan Institute of Technology, Rashtrasant Tukadoji Maharaj Nagpur University, Nagpur, motivated and helped us in bringing out this book.

Dr. B. A. Bhanvase and Dr. R. P. Ugwekar would like to acknowledge Rashtrasant Tukadoji Maharaj Nagpur University officials, various research laboratories, and the advisory and working committee for their support and encouragement from time to time. We would like to thanks all the contributors and their respective organizations.

Dr. Bharat A. Bhanvase
Dr. Rajendra P. Ugwekar
Dr. Raju B. Mankar

INTRODUCTION

This book *Novel Water Treatment and Separation Methods: Simulation of Chemical Processes* is divided into four parts, such as Selected Wastewater Treatment Examples, Separation Techniques, Modeling and Simulation of Chemical Processes, and some experimental reviews that contain the selected articles from the National Conference on Recent Trends in Chemical Engineering and Technology (REACT 2016), held at Laxminarayan Institute of Technology, Rashtrasant Tukadoji Maharaj Nagpur University, Nagpur, in 2016.

The first part contains 11 chapters covering some selected examples in wastewater treatment. In this part, we highlight some recent examples of processes for wastewater treatment such as photocatalytic degradation, emulsion liquid membrane, novel photocatalyst for degradation of various pollutants, and adsorption of heavy metals. This will give better options to readers as this part contains recent development in wastewater treatment.

The second part of this book includes three chapters related to novel separation techniques, such as microwave-assisted extraction, anhydrous ethanol through molecular sieve dehydration, batch extraction from leaves of *Syzygium cumini,* and reactive extraction. These novel separation techniques proved to be advantageous over the conventional separation techniques.

Part three of this book consists of seven chapters related to modeling and simulation of chemical processes. This part includes chapters on flow characteristics of novel solid–liquid multistage circulating fluidized bed, mathematical modeling and simulation of gasketed plate heat exchanger, optimization of adsorption capacity of prepared activated carbon, modeling of ethanol/water separation by pervaporation, and topics on simulation using CHEMCAD software. In this section, we have highlighted the modeling and simulation of various chemical processes.

Finally, in part four of this book, we have included some experimental reviews for the benefits of the readers.

We believe that these diverse chapters included in this book will encourage new ideas, methods, and applications in ongoing advances in this growing area of chemical engineering and technology. Also, we believe that this book will give new insight in the area of wastewater treatment, separation techniques, and modeling and simulation of chemical processes to benefit students, researchers and faculties and industrialists contemporaneously. The book's editors have compiled the articles, written by various authors, included in this book.

PART I
Modern Techniques in Water Treatment

CHAPTER 1

PHOTOCATALYTIC DEGRADATION OF HERBICIDE BY USING AEROXIDE®P-90 TiO$_2$ PHOTOCATALYST AND PHOTO-FENTON PROCESS IN THE PRESENCE OF ARTIFICIAL AND SOLAR RADIATION

V. SAKHALKAR[1], A. KHANDARE[1], M. P. DEOSARKAR[2], and S. P. KAMBLE[1,*]

[1]*Chemical Engineering and Process Development Division, CSIR-National Chemical Laboratory, Dr. Homi Bhabha Road, Pune 411008, Maharashtra, India*

[2]*Chemical Engineering Department, Vishwakarma Institute of Technology, Upper Indira Nagar, Bibvewadi, Pune 411037, Maharashtra, India*

*Corresponding author. E-mail: sp.kamble@ncl.res.in

CONTENTS

ABSTRACT

The presence of herbicide residues in the aquatic environment is an emerging issue due to their uncontrolled release through water and accumulation in the environment that may affect living organisms, environment, and public health. Consequently, the efforts are being made to develop the viable methods to eliminate the herbicides from the environment. The degradation of herbicide particularly amitrole has been investigated in aqueous solutions by using artificial and solar radiation. The effect of Aeroxide TiO_2 P-90 photocatalyst loading (1–4 g L^{-1}) on photocatalytic degradation of amitrole (50 mg L^{-1}) was investigated. The degradation of amitrole (initial concentration ranging from 20 to 100 mg L^{-1}) also studied by using photo-Fenton process ($FeSO_4$ (10–30 mg L^{-1}) and H_2O_2 (100–300 mg L^{-1})). The effect of pH (3–11), effect of co-existing salts (ammonium sulfate, sodium chloride, and sodium carbonate) as well as effect of type of radiation (artificial or solar radiation) on the degradation of amitrole was evaluated and optimized. The degradation of amitrole was accompanied by formation of intermediates which were detected by LC-MS. According to the results presented in this study, the photo-Fenton process using solar radiation is an efficient technique for degradation of amitrole.

1.1 INTRODUCTION

Water pollution due to pesticides is a grave environmental crisis due to their possible toxicity and bioaccumulation. The pesticide compounds are persistent in aquatic environment and can potentially cause adverse health effects. The presence of pesticides in surface and ground water has been recognized as a major problem in many countries. Therefore, the removal of pesticides from aqueous streams has been an issue of immense environmental significance in recent years.[1] The complete degradation of organic pollutants is not possible by conventional approaches such as physicochemical treatment, anaerobic digestion, and activated sludge digestion; as they only relocate the contaminants from one phase to another.[2,3] However, advanced oxidation processes (AOPs), which exploit hydroxyl radicals for environmental remediation, have been effectively working for the degradation of organic compounds including organophosphates.[2-6] AOPs include photo-Fenton, ozone treatment such as O_3/UV, O_3/H_2O_2, photocatalytic and photochemical processes, etc. Photocatalytic and photo-Fenton processes have many advantages like (a) this process occurs at ambient conditions, (b) leads to the absolute mineralization of organic carbon into CO_2 as its final product, and (c) cost effective when sunlight is used as the source of irradiation.[6,7]

Amitrole is widely used for weed control in agriculture and also along roadsides and railways. Amitrole is highly soluble in water and hence frequently found in surface water which can contribute to underground water contamination via leaching.[8,9] For amitrole, some approaches have been explored, including electro-Fenton, activated carbon and nano-filtration; however, these processes have many drawbacks.[10–13] Therefore, in this chapter, photo-Fenton process and photocatalytic process are chosen as promising approaches for the degradation of amitrole. There are no reports on degradation of amitrole by using photo-Fenton in the presence of solar radiation. In this chapter, a comparative study was performed studying the degradation of amitrole using photo-Fenton and photocatalytic processes. The consequence of operating parameters, such as initial concentration of amitrole, $FeSO_4/H_2O_2$ concentration, pH, and presence of co-existing ions on degradation of pesticide by photo-Fenton process was investigated in detail. The experimental conditions were optimized and the intermediates formed during the degradation of amitrole were recognized by using liquid chromatography-mass spectrometry/mass spectrometry (LC-MS/MS) technique.

1.2 MATERIALS AND METHODS

1.2.1 MATERIALS

Aeroxide® P-90 TiO_2 (70:30% w/w anatase to rutile) average particle size of 14 nm and Brunauer–Emmett–Teller (BET) surface area about 70–110 m^2 g^{-1} was the photocatalyst used for degradation. The amitrole, acetonitrile, Na_2CO_3, NaCl, $(NH_4)_2SO_4$, $FeSO_4 \cdot 7 \cdot H_2O$, and H_2O_2 (50% w/v) used in the study were purchased from Merck India Pvt. Ltd, Mumbai.

1.2.2 ANALYTICAL METHODS

The amitrole concentration in the samples was monitored by high performance liquid chromatography (HPLC) made by Dionex (P680) with a UV detector (Dionex UVD 170U). A C-18 column, silica gel (Bischoff, C-18, 5 μm particle size, 250 × 4.6 mm.) was used. The mobile phase was 70% de-ionized water, 30% acetonitrile (isocratic mixture) illuminated at a wavelength of 212 nm. The flow rate of mobile phase was 1 mL min⁻¹. The HPLC

column temperature was maintained at 35°C. In order to separate the catalyst particles, samples were centrifuged and filtered with a polyethersulfone
membrane filter with pore size of 0.45 μm before the HPLC analysis.

1.2.3 DEGRADATION EXPERIMENTS

Experimental setups used for photocatalytic degradation as well as photo-
Fenton process are shown in Figures 1.1a and 1.1b. A visible tungsten lamp
of power 400 W was used as the source of artificial radiation. A known

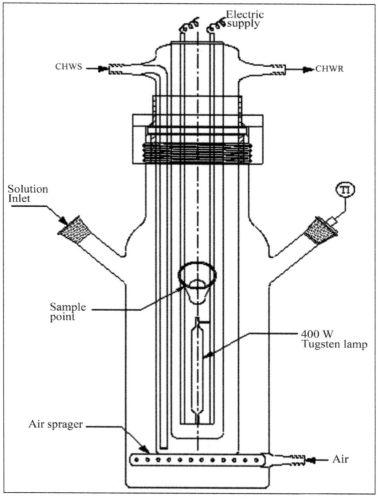

FIGURE 1.1a   Experimental setup of degradation of amitrole using artificial radiation.

concentration of amitrole solution was prepared in the de-ionized water. The temperature of reaction mixture was maintained at room temperature by circulating cold water through the annular space between the lamp and the reactor which can be seen in the diagram below, via a chiller (Julabo FP-50 MA). In every experiment, 500 mL of aqueous solution of amitrole was loaded in the reactor. A ring sparger was used to sparge air at the rate of 1 L min^{-1}, which was situated at the bottom of the reactor for the oxidation of the reaction and also to keep the catalyst in suspension. The photocatalytic degradation experiments were performed using Aeroxide® P-90 TiO$_2$ as the photocatalyst while FeSO$_4$/H$_2$O$_2$ was used as catalyst in the photo-Fenton process. The liquid samples were taken for every 10 min in all experiments from the reactor for concentration monitoring of amitrole and stored in amber-colored bottles to avoid exposure to sunlight. The pH was not monitored during the experimentation.

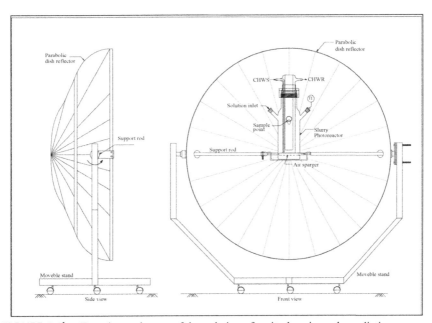

FIGURE 1.1b   Experimental setup of degradation of amitrole using solar radiation.

Similar experimental conditions were used for photo-Fenton degradation of amitrole using solar radiation. A curved (parabolic) reflector with a surface area of 1.5 m^2 was used to focus the solar radiation forming a continuous concentric glowing band of light surrounding the wall of the reactor. After periodic interval of 20 min, the position of the reflector was adjusted with respect to the sun so as to keep constant intensity of incident light. The degradation experiments were performed in the month of January–April; during this period, sky is clear in Pune, India. Normal (direct) solar intensity was measured in $W \cdot m^{-2}$ by "daystar meter" (Daystar Inc., Las Cruces, NM, USA) working on the photocell principle. Average solar radiation was found to be ~850 ± 10 W/m^2.

1.3   RESULTS AND DISCUSSION

1.3.1   COMPARISON OF PHOTOCATALYTIC AND PHOTO-FENTON PROCESS FOR DEGRADATION OF AMITROLE

The degradation of amitrole was performed by photocatalytic and photo-Fenton processes by using solar as the source of radiation. The batches were conducted using 100 mg/L as amitrole initial concentration. Figure 1.2 shows the comparison of degradation of amitrole by photocatalytic and photo-Fenton processes; the rate of degradation of amitrole was higher using photo-Fenton process than that of photocatalytic process. About 40%

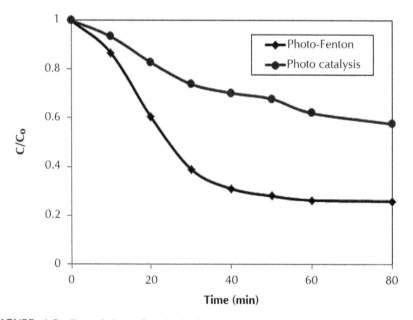

FIGURE 1.2 Degradation of amitrole by photocatalytic and photo-Fenton process (concentration of amitrole = 100 mg/L; $FeSO_4$:H_2O_2 = 10:100 mg/L; Aeroxide® TiO_2 P-90 Dosage = 1 g/L; solar radiation = 840 W/m^2).

degradation of amitrole was observed in case of photocatalytic degradation by using Aeroxide® P-90 TiO_2 photocatalyst while 80% degradation of
amitrole was achieved in the case of photo-Fenton process in a time period
of 80 min. This shows that the photo-Fenton process is more efficient as
compared to the photocatalytic process. Hence, all the remaining batches
were performed using photo-Fenton process.

1.3.2 COMPARISON OF DEGRADATION OF AMITROLE BY USING PHOTO-FENTON PROCESS IN THE PRESENCE OF SOLAR AND ARTIFICIAL RADIATION

Degradation experiments were conducted using 100 mg/L initial concentration of amitrole. The experiments were performed using both artificial radiation (400 W) and solar radiation (Avg. plain solar intensity 850 ± 10 Wm^2).
Figure 1.3 shows the comparison of the rate of degradation of amitrole
using solar and artificial radiation. The rate of degradation of amitrole in the
presence of solar radiation is greater than that of artifical radiation. This is
because the intensity of solar radiation is much higher as compared to that of
the artificial radiation provided. Another reason is that the fraction of useful

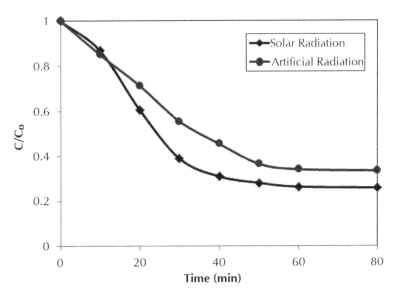

FIGURE 1.3 Degradation of amitrole by photo-Fenton process using artificial radiation and
solar radiation ($FeSO_4$:H_2O_2 = 10:100 mg/L; initial concentration of amitrole = 100 mg/L;
solar = 840 W/m^2; artificial = 400 W).

radiation present in sunlight (in the UV and visible spectra) is substantially higher than that present in artificial radiation. Therefore, further studies were carried out by using solar radiation.

The regions such as South America, South Africa, India, etc. which are in the equitorial region have abundunt and strong sunlight throughout the year, where the treatment of wastewater using solar radiation can have promising results.

1.3.3 OPTIMIZATION OF PROCESS PARAMETERS FOR DEGRADATION OF AMITROLE BY PHOTO-FENTON PROCESS USING SOLAR RADIATION

1.3.3.1 FESO$_4$/H$_2$O$_2$ CONCENTRATION ON DEGRADATION OF AMITROLE

The effect of initial loading of hydrogen peroxide and $FeSO_4$ on the photo-Fenton oxidation of amitrole in aqueous solution was investigated. The quantities of $FeSO_4$:H_2O_2 were varied as 10:100, 20:200, 25:250, and 30:300 mg/L. The ratio of $FeSO_4$:H_2O_2 was kept as 1:10 so that sludge formation could be avoided. As seen from the Figure 1.4, the optimum $FeSO_4$:H_2O_2 ratio was observed at 25:250 mg/L.

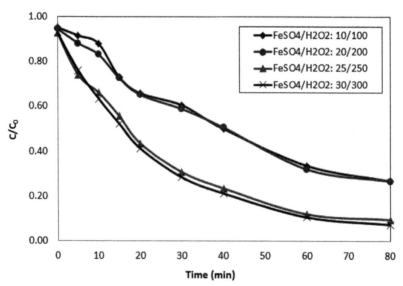

FIGURE 1.4   Effect of $FeSO_4$/H$_2$O$_2$ loading on degradation of amitrole under solar radiation (initial concentration of amitrole = 50 mg/L).

It can be seen from the results that when $FeSO_4$:H_2O_2 concentration ratio is increased, the rate of degradation of amitrole increases. However, the rate of degradation of amitrole at $FeSO_4$:H_2O_2/25:250 mg/L and $FeSO_4$:H_2O_2/30:300 mg/L concentrations was nearly same. This shows that dosage of $FeSO_4$:H_2O_2 at 25:250 mg/L is optimum ratio for degradation of amitrole. At higher $FeSO_4$:H_2O_2 ratio, the sludge formation will be more and the effective utilization of hydrogen peroxide will be less.

1.3.3.2 INITIAL CONCENTRATION OF AMITROLE

The initial concentration of the contaminant is one of the important parameters which disturbs the rate of its degradation. Degradation studies were conducted out using 20, 50, and 100 mg/L initial concentration of amitrole, while the other parameters were not varied. Figure 1.5 shows the effect of original concentration of amitrole on the rate of degradation; it was found that as the original concentration of amitrole increases, the rate of degradation declines. The rate constant (k) decreased from 0.043 to 0.027 min^{-1} with an increase in the amitrole concentration from 20 to 100 mg/L.

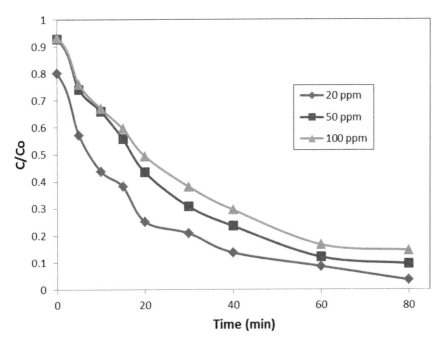

FIGURE 1.5 Effect of initial concentration on degradation of amitrole using solar radiation ($FeSO_4$:H_2O_2 = 25:250 mg/L).

1.3.3.3 pH

The wastewater upcoming from diverse sources has different pH; hence, rate of degradation at different pH was studied. Figure 1.6 shows the rate of deprivation of amitrole at diverse pH. The degradation of amitrole is dependent not only on $FeSO_4$ concentration but also on pH of amitrole solution. The effect of pH on the rate of degradation of amitrole was studied at pH values ranging from 3 to 11. In the basic range, the pH was changed using aqueous 0.1 N NaOH, whereas, in the acidic range, pH was changed using aqueous 0.1 N HCl. Initial concentration of amitrole was 50 mg/L, and $FeSO_4$:H_2O_2 concentration ratio of 25:250 mg/L was used in all these experiments. It was observed that rate of degradation of amitrole was highest at pH 3. Above pH 3, Fe^{3+} starts to precipitate. At pH less than 2, the fraction of Fe (III) in the form of Fe (III)-peroxy complexes becomes very minor and Fe^{2+} formation rate declines.

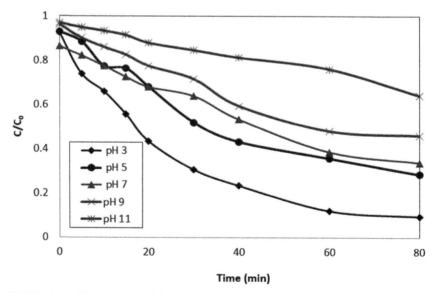

FIGURE 1.6   Effect of pH variation on degradation of amitrole using solar radiation (initial concentration of amitrole = 50 mg/L, and $FeSO_4$:H_2O_2 = 25:250 mg/L).

1.3.3.4 CO-EXISTING IONS

The wastewater contains different salts at different concentration levels separately from the contaminants. These salts have different effects on the degradation, and therefore, it is essential to study the effect of co-existing ions on degradation of amitrole. The effect of anions such as chloride, carbonate, and sulfate were examined using 500 mg/L aqueous solution of their sodium salts (Fig. 1.7). The original concentration of amitrole was 50 mg/L and $FeSO_4:H_2O_2$ concentration was 25:250 mg/L in these experiments. Figure 1.7 indicates that the ammonium sulfate and sodium carbonate salts have a considerably negative effect on the degradation of amitrole. In presence of sulfate and carbonate salts, pH of the solution becomes alkaline, while in the presence of chloride salts, the pH of solution becomes acidic. From the pH study (Section 1.3.3.3), it was clear that higher degradation of amitrole was observed at acidic pH. Hence, sodium chloride salt has positive effect on degradation of amitrole.

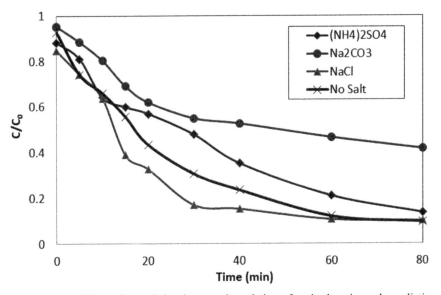

FIGURE 1.7 Effect of co-existing ions on degradation of amitrole using solar radiation (initial concentration of amitrole = 50 mg/L, and $FeSO_4:H_2O_2$ = 25:250 mg/L).

1.4 MECHANISM OF PHOTOCATALYTIC DEGRADATION OF AMITROLE

The pathway of degradation of organic compounds takes place through the hit of OH• on the substrate. The hydroxyl radicals are formed by different reactions ((1.1)–(1.6)).

$$Fe^{2+} + H_2O_2 \rightarrow HO^\bullet + Fe^{3+} + OH^- \qquad (1.1)$$

$$Fe^{3+} + H_2O_2 \rightarrow Fe^{2+} + HO_2^\bullet + H^+ \qquad (1.2)$$

$$Fe^{2+} + HO^\bullet \rightarrow Fe^{3+} + OH^- \qquad (1.3)$$

Photolysis of H_2O_2 and reduction of Fe^{3+} ions under UV light are additional sources of OH radicals which should be considered in photo-Fenton type processes,

$$H_2O_2 + hv \rightarrow HO^\bullet + HO^\bullet \qquad (1.4)$$

$$Fe^{3+} + H_2O + hv \rightarrow Fe^{2+} + HO^\bullet + H^+ \qquad (1.5)$$

$$Amitrole + HO^\bullet \rightarrow Oxidation\ Product \qquad (1.6)$$

The study of pathway of degradation of contaminant is very significant because throughout the degradation of aromatic compounds, several intermediates or byproducts are formed and these may be more harmful than the original compounds.

The intermediates formed during the photo-Fenton process of amitrole were analyzed using UPLC-MS/MS. Two intermediates were detected and their respective molecular masses were determined by using UPLC-MS/MS. The structures and retention time of intermediates are summarized in Table 1.1. Analogous intermediates were observed during the degradation of amitrole by Catastini et al.[11,12] In the case of amitrole, oxidations by hydroxyl

TABLE 1.1 Intermediates Formed during the Photo Degradation of Amitrole.

Sr.	RT (min)	[M+h]⁺ and molecular formula	Structure
1.	0.36	84 $C_2H_4N_4$	
2.	0.43	101 $C_2H_3N_3O_2$	
3.	0.43	100 $C_2H_4ON_4$	

radicals may lead to the formation of adduct OH$^\bullet$-amitrole. In the presence of oxygen, the adduct OH$^\bullet$-amitrole may form 5-hydroxyl amitrole and urazol.

1.5 KINETICS OF DEGRADATION OF AMITROLE

The kinetics of degradation reaction was studied by monitoring the change in concentration of amitrole on HPLC during photo-Fenton process under solar radiation (Fig. 1.8). The plot between $-\ln(C/C_o)$ and time (t) follows first order kinetics. As per reaction (1.6) the concentration of hydroxyl radicals is in excess with respect to amitrole. This concludes that the degradation follows pseudo first order rate kinetics, given in eq 1.7. The rate constant, k (min^{-1}) decreased from 0.043 to 0.027 min^{-1} with increasing concentration from 20 to 100 mg/L, respectively.

$$\ln\left(\frac{C}{C_o}\right) = -k \times t \tag{1.7}$$

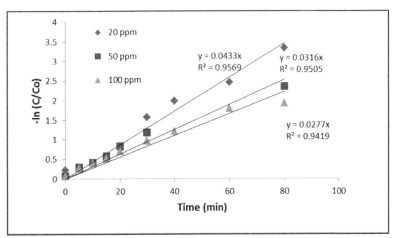

FIGURE 1.8 Kinetic parameters at different initial concentrations of amitrole under solar radiation (FeSO$_4$:H$_2$O$_2$/25:250 mg/L).

1.6 CONCLUSION

The degradation of amitrole was studied by photocatalytic and photo-Fenton processes and it was found that photo-Fenton process was more competent for complete degradation of amitrole. The consequence of various working parameters on degradation of amitrole was studied and optimum parameters were established. The rate of degradation of amitrole was highest at pH 3. The effect of sodium carbonate, sodium chloride, and ammonium sulfate on degradation of amitrole was examined and it was concluded that the carbonate salts have a considerably harmful effect on the degradation of amitrole. However, ammonium sulfate and sodium chloride have no noteworthy effect on the rate of degradation. Amitrole degradation takes place with the formation of intermediate organic compounds like urazol, 5-hydroxyamitrole. The degradation of amitrole by photo-Fenton process follows pseudo first-order kinetics. Thus, it can be concluded that photo-Fenton process is a promising approach for the treatment of pesticide-containing wastewater.

ACKNOWLEDGMENTS

Authors are thankful for the financial support provided by CSIR, Government of India, under the project "Development of Sustainable Waste Management Technologies for Chemical and Allied industries (SETCA) Project # CSC-0113".

KEYWORDS

- amitrole
- photocatalysis
- photo-Fenton
- Aeroxide P-90 TiO_2 photocatalyst
- solar radiation
- kinetics of photocatalytic degradation

REFERENCES

1. Dhiraj, S.; Paramjeet, K. Heterogeneous Photocatalytic Degradation of Selected Organophosphate Pesticides: A Review. *Crit. Rev. Environ. Sci. Technol.* **2012,** *42,* 2365–2407.

2. Malato, S.; Blanco, J.; Cáceres, J.; Fernández-Alba, A. R.; Agüera, A.; Rodrıguez, A. Photocatalytic Treatment of Water-Soluble Pesticides by Photo-Fenton and TiO_2 Using Solar Energy. *Catal. Today*. **2002**, *76,* 209–220.

3. Oturan, M. A.; Edelahi, M. C.; Oturan, N.; Kacemi, K. E.; Aaron, J. J. Kinetics of Oxidative Degradation/Mineralization Pathways of the Phenyl Urea Herbicides Diuron, Monuron and Fenuron in Water during Application of the Electro-Fenton Process. *Appl. Catal. B*. **2010**, *97,* 82–89.

4. Maldonado, M. I.; Passarinho, P. C.; Oller, I.; Gernjak, W.; Fern´andez, P.; Blanco, J.; Malato, S. Photocatalytic Degradation of EU Priority Substances: A Comparison between TiO_2 and Fenton Plus Photo-Fenton in a Solar Pilot Plant. *J. Photochem. Photobiol. A Chem*. **2007**, *185,* 354–363.

5. Lapertot, M.; Ebrahimi, S.; Dazio, S.; Rubinelli, A.; Pulgarin, C. Photo-Fenton and Biological Integrated Process for Degradation of Mixtures of Pesticides. *J. Photochem. Photobiol. A Chem*. **2007**, *186,* 34–40.

6. Shetty, R.; Chavan, V. B.; Kulkarni, P. S.; Kulkarni, B. D.; Kamble, S. P. Photocatalytic Degradation of Pharmaceuticals Pollutants Using N-Doped TiO_2 Photocatalyst: Identification of CFX Degradation Intermediates. *Indian Chem. Eng*. **2016**, In Press. DOI: 10.1080/00194506.2016.1150794.

7. Bhatkhande, D. S.; Pangarkar, V. G.; Beenackers, A. A. Photocatalytic Degradation for Environmental Applications – A Review. *J. Chem. Technol. Biotechnol*. **2001**, *77,* 102–116.

8. Sun, Y.; Peng, P. F.; Wang, D.; Li, J. Q.; Cao, Y. S. Determination of Amitrole in Environmental Water Samples with Precolumn Derivatization by High-Performance Liquid Chromatography. *J. Agric. Food Chem*. **2009**, *57,* 4540–4544.

9. Chicharro, M.; Zapardiel, A.; Bermejo, E.; Moreno, M. Determination of 3-Amino-1, 2, 4- Triazole (Amitrole) in Environmental Waters by Capillary Electrophoresis. *Talanta*. **2003**, *59,* 37–/45.

10. Fontecha-Cámara, M. A.; Álvarez-Merino, M. A.; Carrasco-Marín, F.; LópezRamón, M. V.; Moreno-Castilla, C. Heterogeneous and Homogeneous Fenton Processes Using Activated Carbon for the Removal of the Herbicide Amitrole from Water. *Appl. Catal. B*. **2011**, *101,* 425–430.

11. Moreno-Castilla, C.; Fontecha-Cámara, M. A.; Álvarez-Merino, M. A.; LópeRamó, M. V.; Carrasco-Marín, F. Activated Carbon Cloth as Adsorbent and Oxidation Catalyst for the Removal of Amitrole from Aqueous Solution. *Adsorption*. **2011**, *17,* 413–419.

12. Catastini, C.; Rafqah, S.; Mailhot, G.; Sarakha, M. Degradation of Amitrole by Excitation of Iron(III) Aqua Complexes in Aqueous Solutions. *J. Photochem. Photobiol. A Chem*. **2004**, *162,* 97–103.

13. Fontecha-Ca´mara, M. A.; Lo´pez-Ramo´n, M. V.; Lvarez-Merino, M. V. V.; MorenCastilla, C. Effect of Surface Chemistry, Solution pH, and Ionic Strength on the Removal of Herbicides Diuron and Amitrole from Water by an Activated Carbon Fiber. *Langmuir*. **2007**, *23,* 1242–1247.

CHAPTER 2

STUDIES ON REMOVAL OF ANIONIC DYE USING EMULSION LIQUID MEMBRANE

R. KANKATE[1], S. GHODKE[1,*], and S. SONAWANE[2]

[1]*Department of Chemical Engineering, Sinhgad College of Engineering, Pune 41104, Maharashtra, India*

[2]*Department of Chemical Engineering, National Institute of Technology, Warangal 506004, Andhra Pradesh, India*

*Corresponding author. Email: saghodke@gmail.com

CONTENTS

ABSTRACT

Emulsion liquid membrane (ELM) separation technique is found to be an efficient methodology for the extraction of complex textile dyes from industrial wastewater. Various key parameters need to be studied for the high selectivity and stability of the membrane for extraction process. In this study, we report the extraction of Acid Red 1 (AR) dye using ELM. The membrane prepared was found to have nanometric dimensions. The concentration of dye solution was measured using UV–visible spectrophotometer. Effects of operating parameters such as membrane synthesis methods, contact time, surfactant concentration, types of internal phase, etc. were studied in detail. The method is found to be efficient in successful removal of dye from the aqueous solution.

2.1   INTRODUCTION

Large developments in the fields of process, pharmaceutical, fertilizer, textile, tannery, automobile, and allied industries have posed problems related to effluent treatments.[1,2] Consequently, problems related to formation of complex and stable products, intermediates, and by-products have been of interest to the various environmental scientists. Dyes are complex compounds utilized to deposit on the surfaces of the textile products and impart color. Contamination of ground and potable with dye effluent produces significant impact limiting its use for the domestic purposes like cooking, bathing, drinking, etc. This is because of significant developments in dye molecules and pigments which may produce substantial amount of coloring effect for a small concentration ranges (1 mg/L).[3-5] Dye wastewater is not only carcinogenic but also produces a layer onto water surfaces restricting sunlight to pass through. Due to such detrimental effects, pollution control agencies have provided stringent norms on dye wastewater effluents from various industries. Considering very large amount of dye wastewater thrown out of industries, the treatment methodologies need to be effective, economical, and regenerative.[6-9] Commonest methodologies practiced are physical (e.g., adsorption, membrane, coagulation, etc.), chemical (oxidative, photochemical, electrochemical, etc.), and biological treatments (microbial biomass, anaerobic bioremediation, etc.). However, such methods are found to be ineffective due to one or more disadvantages like expensive, sludge producing, by-product formation, unreliability, and control parameters.[5-7,10]

There are three basic types of dyes: anionic, cationic, and non-ionic dyes.[11] Presence of sulfonate (SO_3^-) groups produces a net negative charge onto the anionic dye; similarly, sulfur-containing groups or protonated amines produce positive charge onto cationic dye.[12] Liquid membranes like bulk liquid membrane (BLM),[13,14] supported liquid membrane (SLM),[15] and emulsion liquid membrane (ELM)[16] have been found to be effective in treatment of dye wastewaters. BLM and SLM are found to suffer from smaller interfacial area and loss of organic phase. Out of these three methods, ELM has been effective in removing pollutants at very low concentration levels. Many successful attempts of ELM have been reported to remove phenolic compounds,[17] aniline,[18] heavy metals,[19] organic compound,[20] amino acid,[21] etc. ELM is found to be advantageous in possessing higher interfacial area, relatively lower cost, and regenerative. Internal phase is surrounded by membrane phase in the form of small droplets which is dispersed in external aqueous solution. External phase, membrane phase, and internal phase combine to form emulsion membrane. During solute transport, solute first moves through external phase to membrane phase in the molecular state or in the form of complex with a carrier.[20,22] Further, it converts to ions when it reaches the surface between membrane and internal phase. Low stability due to membrane rupture resulting in loss in internal phase is the major limitation for practical applications of ELM.[23] Careful analysis and study on factors affecting membrane stability is required to maintain the membrane stability. On the other hand, higher membrane stability may lead to the difficulty in demulsification process, which in turn limits the membrane applications.[20]

Considering the importance of emulsion preparation, various methods like stirrer, colloid mill, mixer, valve homogenizer, and ultrasound generation have been used.[24] During ultrasonic cavitation, small cavities form, grow, and collapse resulting in very high local energy density.[25] Because of rapid dissociation of vapors trapped in the cavitating bubbles, free radicals are produced in the process, which results in alteration of reaction mechanisms or chemical reaction intensification.[26,27] Therefore, ultrasound acoustic cavitation is found to produce useful chemical and physical changes during chemical processes. As per the various studies mentioned in the literature,[24,28,29] ultrasound is able to produce smaller droplet size for a given emulsion as compared to conventional stirring. The emulsion droplets synthesized under ultrasound are able to provide narrow size distribution in the order of few micrometers.[30] Meriem Djenouhat et al. discussed about the use of ultrasound to create ELM and the emphasis was to study the membrane stability in a short range of low power 5–20 W.[31] The membrane

2.2.2.1   ULTRASOUND-ASSISTED MEMBRANE PREPARATION

During emulsion (W/O) membrane synthesis, internal phase containing 0.5 N aqueous solution of sodium carbonate was added drop by drop into a beaker containing organic phase. The organic phase prepared contains (w/w) span 80 4%, D2EHPA 10%, and diluent 86%. The ratio of internal phase to the organic phase was 1:1 to make the mixture equal to 50 mL. For sonication procedure, a 20-mm dia ultrasound probe was placed at the interface for 13 min (ON time 60 s and OFF time 5 s). The coolant flow in reactor jacket was maintaining the reactor temperature below 25°C. This results in a continuous membrane phase used for further extraction procedure.

2.2.2.2   EXTRACTION USING ELM

During dye extraction or stripping phase, the W/O emulsion synthesized under ultrasound was added to a feed solution containing dye under constant stirring using magnetic stirrer. In a typical extraction experiment, about 50 mL of emulsion was added to 200 mL aqueous dye (feed) solution and the mixture was stirred at 500 rpm. The solution pH was monitored to identify the membrane rupture, that is, decrease in solution pH indicates the dripping of H^+ ions in the external phase. Each membrane preparation and extraction procedure was repeated thrice and the percentage deviation was ±3.

The amount of dye extracted was reported in the form of extraction efficiency. The extraction efficiency was estimated using following relation

$$\text{Extraction efficiency (\%)} = \frac{C_o - C}{C_o} \times 100 \qquad (2.1)$$

where C_o is the initial concentration of dye in the feed phase (ppm);
C is the concentration of the dye in the feed phase at time t.

2.2.2.3   CHARACTERIZATION

All the membranes produced during experimentation were analyzed for the emulsion size using particle size analyzer (Make—Horiba, Japan, Model—SZ 100). Dye solution concentration was monitored using UV spectrophotometer (Shimadzu, Model-UV 2600). The maximum absorbance (λ_{max}) obtained for AR 1 dye was 531 nm. All the dye samples were centrifuged at 2000 rpm using centrifuge (Make—REMI, Model—R-BC). The solution pH was measured using Bench Top pH Meter (Model-PH 600R).

2.3   RESULTS AND DISCUSSIONS

2.3.1   MEMBRANE PREPARATION METHOD

One of the most important criteria of extraction using ELM is extraction efficiency. A typical W/O emulsion can be prepared by simple stirring. Higher speeds of stirring are required to decrease the emulsion size which in turn yields to higher extraction efficiency. High-speed stirring may lead to the emulsion breakdown causing lesser extraction efficiency. Considering this, membrane preparation was studied under magnetic stirring and ultrasound. For this, 4% span 80 and D2EHPA 10% were added to hexane and the mixture was stirred at 500 rpm for 13 min. Similar compositions were sonicated for 13 min (probe dia—20 mm, power—60 W, frequency—22.5 kHz). The cycles for sonication were ON time 60 s and OFF time 5 s and the corresponding extraction efficiency for such emulsions was measured. Figure 2.1 indicates the extraction efficiency of emulsions prepared under ultrasound and magnetic stirring. The curve indicating extraction efficiency under ultrasound indicated higher efficiency as compared to magnetic stirring. Under ultrasound, the 85% dye was extracted within first 3 min which subsequently reached to 96% after 30 min. At the same conditions, emulsion

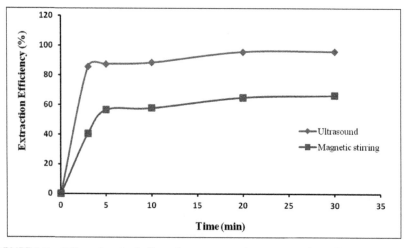

FIGURE 2.1   Effect of method of membrane preparation on extraction efficiency.

from magnetic stirring could produce only 40% in first 3 min which subsequently reached to 66% after 30 min. These higher extraction rates for ultrasound prepared emulsions are attributed to the smaller emulsion droplets producing higher surface area which is comparably smaller in emulsion under magnetic stirring.

2.3.2 EFFECT OF SONICATION TIME

Success of any membrane preparation depends on the extent of dispersion of internal phase into the membrane phase. Higher ultrasound power provided to create emulsions decreases the emulsion breakage and provides smaller emulsion size.[30] This may be because more time is required to coalesce the smaller droplets which provides higher membrane stability. Membrane was prepared by adding 4% span 80 and D2EHPA 10% to hexane and the mixture was sonicated at different sonication times from 6 to 27 min. The membrane droplet size (Z average) at various sonication times is represented in Figure 2.2. It is observed that with increase in sonication time, the $Z_{average}$ of emulsion droplet increases; initially, it is very large where emulsion was yet to form. At 13 min, $Z_{average}$ was found to be less, which is acceptable. After 13 min, sample particle coagulation occurs in the emulsion and $Z_{average}$ increases. Table 2.1 indicates the values of $Z_{average}$ corresponding to each sonication time. From the table, it is observed that value of $Z_{average}$ at 13 min

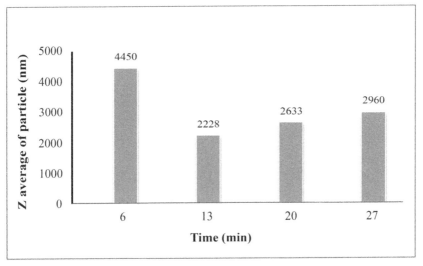

FIGURE 2.2   Effect of sonication time on to $Z_{average}$ of emulsion membrane.

is comparatively less than its values at any other sonication time. Hence, all the experimentations for membrane preparation were conducted for 13 min sonication time.

TABLE 2.1 Values of $Z_{average}$ (nm) at Various Sonication Time.

Particulars	Sonication time (min)			
	6	13	20	27
$Z_{average}$ (nm)	4450.1	2228.2	2633.6	2961.6

2.3.3 EFFECT OF CONTACT TIME

The effect of contact time on increase of extraction efficiency was determined at various contact times. A 20 ppm, feed solution was used to determine the extraction efficiency at different contact time. The graph in Figure 2.3 indicates an increasing trend in the extraction efficiency with increase in the contact time. The results indicated an initial rapid extraction of 66% within first 2 min which increased further to 85% in first 12 min. It further increased with an increase in contact time to almost equilibrium stage of 92%. The increase in percent extraction was due to increase in time allotted for more mass transfer through the membrane phase.

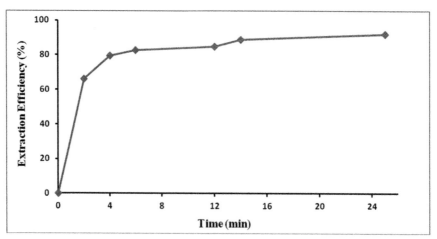

FIGURE 2.3 Effect of contact time on extraction efficiency. Internal phase to organic phase ratio (1:1), hexane 86%, D2EHPA—10%, span 80 4%, membrane phase 50 mL, feed (dye solution) 200 mL, internal phase—Na_2CO_3 0.5 N, stirring speed 500 rpm, and sonication time 13 min.

2.3.4 EFFECT OF SURFACTANT CONCENTRATION

The membrane stability, swelling characteristics, and breakup strongly depend on the surfactant concentration. Figure 2.4 depicts the extraction efficiency at various concentrations of surfactant (span 80). The extraction efficiency was found to increase from 73 to 96% for increase in surfactant from 1 to 4% for first 10 min. On the contrary, further increase in the surfactant concentration could not increase the extraction. This may be due to increase in amount of surfactant at the surface of aqueous and organic phase leading to higher mass transfer resistance for dye molecules. Hence, all further experimentations were conducted at 4% concentration.

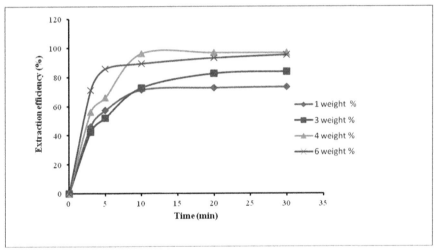

FIGURE 2.4 Effect of surfactant concentration on extraction efficiency. Internal phase to organic phase ratio (1:1), hexane 86%, D2EHPA—10%, span 80 4%, membrane phase 50 mL, feed (dye solution) 200 mL, internal phase—Na_2CO_3 0.5 N, stirring speed 500 rpm, and sonication time 13 min.

2.3.5 EFFECT OF TYPE OF INTERNAL PHASE

Selection of internal phase is vital, as it is responsible for the transport of dye molecules during extraction process. Considering the anionic behavior of dye molecule, different types of basic internal phases were selected. Effect of various types of internal phase is shown in Figure 2.5.

For this purpose, various phases like ammonium hydroxide, sodium hydroxide, and sodium carbonate were used. Greater amount of extraction efficiency was found with sodium carbonate as compared to sodium hydroxide and ammonium hydroxide. With sodium carbonate, 82% extraction efficiency was obtained in first 5 min and 93% in 30 min. This higher extraction efficiency may be due to rapid separation of Na^+ and CO_3^- ions resulting into dye-extractant complex reacting with the Na^+ ions.

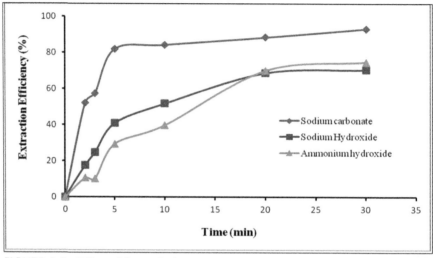

FIGURE 2.5   Effect of internal phase type on extraction efficiency. Internal phase to organic phase ration (1:1), hexane 86%, D2EHPA—10%, span 80 4%, internal phase concentration—0.5 N, membrane phase 50 mL, feed (dye solution) 200 mL, stirring speed 500 rpm, and sonication time 13 min.

2.3.6   TRANSPORT MECHANISM

Figure 2.6 indicates plausible mechanism of transfer of AR dye across liquid membrane. Extractant D2EHPA plays a major role in transporting dye molecules from feed solution into membrane phase. A neutral pair complex is formed from the phosphoryl group of D2EHPA which leads to formation of chelating compounds resulting in the distribution to the organic phase. Ion pair formed diffuses across the membrane and toward the membrane strip solution interface as a result of driving force provided by concentration gradient. Ion-paired dye is released at membrane strip solution interface. Back diffusion of the dye also occurs across the membrane.[30–32]

Here, the dye anion (Dye⁻) reacts with the carrier [(H⁺R)] at the external interface to form Dye D2EHPA complex (HR-Dye).[33]

$$(Dye^-) + (H^+R) \longleftrightarrow (HR\text{-}Dye) \tag{1}$$

Dye extractant complex then diffuses through the membrane phase to the internal phase.

Dye-D2EHPA complex then release through membrane phase to stripping phase and it reacts with the sodium ion provided by internal phase, that is, aqueous solution of sodium carbonate.

$$(HR\text{-}Dye) + Na^+ \longleftrightarrow Dye^- \; Na^+ + (H^+R) \tag{2}$$

$$Dye^- \; Na^+ \longleftrightarrow Dye^- + Na^+ \tag{3}$$

The dye is finally in the receiving phase and the extractant is ready for the cycle.

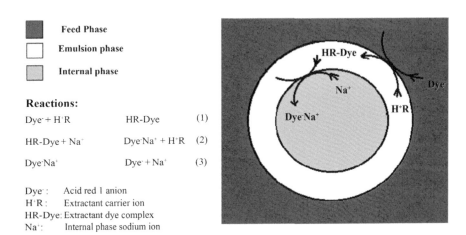

FIGURE 2.6   Plausible mechanism of transfer of Acid Red 1 dye across liquid membrane.

2.4   CONCLUSION

A study of removal of anionic dye (AR) using ELM was undertaken successfully. Use of ultrasound (probe dia—20 mm, power—60 W, and frequency—22.5 kHz) in membrane preparation was found to create smaller emulsion droplets as compared to conventional stirring. The emulsion droplets formed under ultrasound was found to have harmonic mean $Z_{average}$ size of 2228 nm which is within the permissible limit and indicated narrow size distribution. The membrane synthesis under ultrasound showed an excellent stability for a 13 min sonication time. The effect of various parameters such as membrane preparation method, sonication time, contact time, surfactant concentration, and type of internal phase were studied in detail. A result shows that ultrasonication process is more effective for the emulsion formation, as it can form very minute droplet-size particles. The optimum concentration of surfactant was found to be 4%. Basic internal phase Na_2CO_3 provides better results for the extraction of anionic dye than other internal phases. The results of extraction of anionic dye are found to be useful for further practical purposes.

KEYWORDS

- **emulsion liquid membrane**
- **contact time**
- **extraction**
- **separation**
- **ultrasound**
- **anionic dye**

REFERENCES

1. Adegoke, K. A.; Bello, O. S. Dye Sequestration Using Agricultural Wastes as Adsorbents. *Water Res. Ind.* **2015**, *12*, 8–24.
2. Mu, B.; Wang, A. Adsorption of Dyes onto Palygorskite and its Composites: A Review. *J. Environ. Chem. Eng.* **2016**, *4*, 1274–1294.
3. Xian, Z.; Farzana, M. H.; Meenakshi, S. Decolorization and Detoxification of Acid Blue 158 Dye Using Cuttlefish Bone Powder as Co-Adsorbent via Photocatalytic Method. *J. Water Pro. Eng.* **2014**, *2*, 22–30.
4. Tan, K. B.; Vakili, M.; Horri, B. A.; Poh, P. E.; Abdullah, A. Z.; Salamatinia, B. Adsorption of Dyes by Nanomaterials: Recent Developments and Adsorption Mechanisms. *Sep. Purf. Technol.* **2015**, *150*, 229–242.
5. Wang, M.; Wang, L. Synthesis and Characterization of Carboxymethyl Cellulose/ Organic Montmorillonite Nanocomposites and its Adsorption Behavior for Congo Red Dye. *Water Sci. Eng.* **2013**, *6*, 272–282.
6. Sonawane, S.; Chaudhari, P.; Ghodke, S.; Ambade, S.; Mirikar, A.; Bane, A.; Gulig, S. Combined Effect of Ultrasound and Nanoclay on Adsorption of Phenol. *Ultrason. Sonochem.* **2008**, *15*, 1033–1037.

7. Ghodke, S.; Sonawane, S.; Gaikawad, R.; Mohite, K. TiO$_2$/Nanoclay Nanocomposite for Phenol Degradation in Sonophotocatalytic Reactor. *Can. J. Chem. Eng.* **2012,** *90,* 1153–1159.
8. Sivashankar, R.; Sathya, A. B.; Vasantharaj, K.; Sivasubramanian V. Magnetic Composite an Environmental Super Adsorbent for Dye Sequestration – A Review. *Environ. Nanotech. Monitor. Manag.* **2014,** *1–2,* 36–49.
9. Hongqi, S.; Xiaohui, F.; Shaobin, W.; Ming, H.; Moses, O. Combination of Adsorption, Photochemical and Photocatalytic Degradation of Phenol Solution over Supported Zinc Oxide: Effects of Support and Sulphate Oxidant. *Chem. Eng. J.* **2011,** *170,* 270–277.
10. Roya, D.; Majid, M.; Shadi, S.; Horst, B.; Moghadame, M.; Jamal, S. Novel Durable Bio-Photocatalyst Purifiers, a Non-heterogeneous Mechanism: Accelerated Entrapped Dye Degradation into Structural Polysiloxane-Shield Nano-Reactors. *Colloids Surf. B.* **2013,** *101,* 457–464.
11. Mohamad, A.; Dalia, K.; Wan, A.; Wan, A.; Azni, I. Cationic and Anionic Dye Adsorption by Agricultural Solid Wastes: A Comprehensive Review. *Desalination.* **2011,** *280,* 1–13.
12. Emna, E.; Joelle, D.; Fadila, D.; Inès, M.; Amélie, A.; Fabienne, H.; Gilles, M. Efficient Anionic Dye Adsorption on Natural Untreated Clay: Kinetic Study and Thermodynamic Parameters. *Desalination.* **2011,** *275,* 74–81.
13. Rouhollah, K.; Jahan, B.; Farzaneh, S.; Reza, R. Application of Bilinear Least Squares/ Residual Bilinearization in Bulk Liquid Membrane System for Simultaneous Multi-component Quantification of Two Synthetic Dyes. *Chemometr. Intell. Lab.* **2015,** *144,* 48–55.
14. Soniya, M.; Muthuraman, G. Comparative Study between Liquid–Liquid Extraction and Bulk Liquid Membrane for the Removal and Recovery of Methylene Blue from Wastewater. *Ind. Eng. Chem. Res.* **2015,** *30,* 266–273.
15. Grace, M.; Eulsaeng, C.; Arnel, B. Dye/Water Separation through Supported Liquid Membrane Extraction. *Chemosphere.* **2010,** *80,* 894–900.
16. Das, C.; Meha, R.; Gagandeep, A.; Sunando, D.; Sirshendu, D. Removal of Dyes and Their Mixtures from Aqueous Solution Using Liquid Emulsion Membrane. *J. Hazard Mater.* **2008,** *159,* 365–371.
17. Balasubramanian, A.; Venkatesan, S. Removal of Phenolic Compounds from Aqueous Solutions by Emulsion Liquid Membrane Containing Ionic Liquid [BMIM] + [PF6]− in Tributyl Phosphate. *Desalination.* **2012,** *289,* 27–34.
18. Datta, S.; Bhattacharya, P.; Verma, N. Removal of Aniline from Aqueous Solution in a Mixed Flow Reactor Using Emulsion Liquid Membrane. *J. Membr. Sci. Technol.* **2003,** *226,* 185–201.
19. García, M.; Acosta, A.; Marchese, J. Emulsion Liquid Membrane Pertraction of Cr(III) from Aqueous Solutions Using PC-88A as Carrier. *Desalination.* **2013,** *318,* 88–96.
20. Wei, P.; Hongpu, J.; Haochun, S.; Chunjian, X. The Application of Emulsion Liquid Membrane Process and Heat-Induced Demulsification for Removal of Pyridine from Aqueous Solutions. *Desalination.* **2012,** 286, 372–378.
21. Pawel, D.; Piotr, W. Extraction of Amino Acids with Emulsion Liquid Membranes Using Industrial Surfactants and Lecithin as Stabilisers. *J. Membr. Sci. Technol.* **2000,** *172,* 223–232.
22. Roman, M.; Bringas, E.; Ibanez, R.; Ortiz, I. Liquid Membrane Technology: Fundamentals and Review of its Applications. *J. Chem. Technol. Biotech.* **2010,** *85,* 2–10.

23. Dâas, A.; Hamdaoui, O. Extraction of Bisphenol A from Aqueous Solutions by Emulsion Liquid Membrane. *J. Membr. Sci. Technol.* **2010,** *348,* 360–368.
24. Ruey, S.; Kung, H. Ultrasound-Assisted Production of W/O Emulsions in Liquid Surfactant Membrane Processes. *Colloids Surf. A.* **2004,** *238,* 43–49.
25. Suslick, K. The Chemical Effects of Ultrasound. *Sci. Am.* **1990,** *247,* 1439.
26. Gogate, P.; Kabadi, A. M. A Review of Applications of Cavitation in Biochemical Engineering/Biotechnology. *Int. J. Biochem.* **2009,** *44,* 60–72.
27. Sonawane, S.; Chaudhari, P.; Ghodke, S.; Parande, M.; Bhandari, V.; Mishra, S.; Kulkarni, R. Ultrasound Assisted Synthesis of Polyacrylic Acid–Nanoclay Nanocomposite and its Application in Sonosorption Studies of Malachite Green Dye. *Ultrason. Sonochem.* **2009,** *16,* 351–355.
28. Chanamai, R.; Coupland, J.; McClements, D. Effect of Temperature on the Ultrasonic Properties of Oil-in-Water Emulsions. *Colloids Surf. A.* **1998,** *139,* 241–250.
29. Ooi, S.; Biggs, S. Ultrasonic Initiation of Polystyrene Latex Synthesis. *Ultrason. Sonochem.* **2000,** *7,* 125–133.
30. Meriem, D.; Hamdaoui, O.; Chiha, M.; Mohamed, H. Ultrasonication-Assisted Preparation of Water-in-Oil Emulsions and Application to the Removal of Cationic Dyes from Water by Emulsion Liquid Membrane Part 2. Permeation and Stripping. *Separ. Purifi. Technol.* **2008,** *63,* 231–238.
31. Djenouhat, M.; Hamdaoui, O.; Chiha, M.; Mohamed, H. Ultrasonication-Assisted Preparation of Water-in-Oil Emulsions and Application to the Removal of Cationic Dyes from Water by Emulsion Liquid Membrane Part 1: Membrane Stability. *Separ. Purifi. Technol.* **2008,** *62,* 636–641.
32. Muthuraman, G.; Palanivelu, K.; Teng, T. Transport of Cationic Dye by Supported Liquid Membrane Using D2EHPA as the Carrier. *Color. Tech.* **2010,** *126,* 97–102.
33. Othman, N.; Zailani, S.; Mili, N. Recovery of Synthetic Dye from Simulated Wastewater Using Emulsion Liquid Membrane Process Containing Tri-dodecyl Amine as a Mobile Carrier. *J. Hazard Mater.* **2011,** *198,* 103–112.

CHAPTER 3

ZINC OXIDE MICROARCHITECTURES WITH EXPOSED CRYSTAL FACES FOR ENHANCED PHOTOCATALYTIC ACTIVITY

S. P. CHAUDHARI[1], S. P. MESHRAM[2,*], P. D. JOLHE[3], G. N. CHAUDHARI[1,*], and A. B. BODADE[1]

[1]Department of Chemistry, Nanotechnology Research Laboratory, Shri Shivaji Science College, Amravati 444602, Maharashtra, India

[2]Department of Nanoscience and Nanotechnology, Centre for Materials for Electronics Technology, Pune 411008, Maharashtra, India

[3]Department of Biotechnology, Sinhgad College of Engineering, Savitribai Phule Pune University, Pune 411041, Maharashtra, India

*Corresponding author. E-mail: gnchaudhari@gmail.com, satishmeshram99@gmail.com

CONTENTS

ABSTRACT

Herein this chapter, Triton-X mediated facile sonochemical method for the synthesis of hierarchical ZnO microarchitectures is demonstrated. The as-synthesized ZnO architectures are characterized by variety of characterization techniques namely, X-ray diffraction (XRD), field emission scanning electron microscopy (FESEM) coupled with energy dispersive analysis of X-rays (EDAX), ultraviolet–visible spectroscopy (UV–visible), and photoluminescence (PL) spectroscopy. FESEM results revealed that, the as-synthesized ZnO architectures resembles hierarchical flower-like architectures and are built-up of large number of interconnected nanosheets in a regulated fashion which demonstrated improved transportation and diffusion of rhodamine B (RhB) molecules during photocatalytic degradation.

3.1 INTRODUCTION

Recently, remarkable efforts have been devoted to synthesize various nanoscale building blocks with controlled sizes and morphologies such as nanoparticles, nanowires, nanobelts, nanotubes, and nanoflakes.[1–4] Increasing attention has been paid to the synthesis of complex three-dimensional (3D) micro/nanoarchitectures that are assembled using 1D and 2D nanoscale building units. These hierarchical architectures possess unique properties that are distinct from the properties of monomorphological structures.

Zinc oxide (ZnO) (having space group = P_63 mc; $a = 0.32495$ nm, $c = 0.52069$ nm) is an n-type semiconductor with a direct wide band gap (3.37 eV) and large excitation binding energy (60 meV).[5] Zinc oxide has proved extensive applications in diverse fields such as piezoelectric nanogenerators, nanolasers, solar cells, chemical sensors, and photocatalysis.[6–10] Among these, photocatalysis is one of the important applications of ZnO. Various organic pollutants can be effectively degraded over ZnO through the photocatalytic generation of hydrogen peroxide.[11–14] The photocatalytic reactions are believed to proceed at surfaces where nanosized semiconductor can offer increased surface area and thus increased decomposition rate. However, during aging, aggregation of nanometer-scaled building blocks (such as nanoparticles, nanorods, and nanosheets) results in the unwanted reduction in the active surface area. Immobilizing the photocatalysts in the form of a thin film on a substrate may provide an alternative to prevent aggregation.[15,16] In this case, yet the efficiency is not satisfactory compared to that of the suspensions. Moreover, catalytic activities of crystalline materials depend on their exposed crystal faces. For example, anatase TiO$_2$ crystals rich in (001) surfaces possess high catalytic reactivity.[17] TiO$_2$ possessing mixture of anatase and rutile particles, the (110) surfaces provide reductive sites, and (011) surfaces provide oxidative sites.[18] (111) faces of Platinum (Pt) show high methanol oxidation reactivity in fuel cell.[19] (001) surfaces of ZnO have been found to be chemically active whereas (00-1) surfaces were found to be inert.[20,21] Consequently, the synthesis of nanostructured ZnO having high surface area that is stable against aggregation and fine control over exposed crystal faces is crucial for exploring its potential use as a smart material for vivid application. In this regard, several researchers have utilized vapor phase methods or wet chemical processes.[21] In particular, solution-based approaches utilizing organic agents have been successfully used to control the growth of ZnO crystal. The various organic agents that have been employed so far include ethylenediamine,[22] citrates,[23–26] amino acids,[27] polyacrylamide,[28] sodium poly(styrenesulfonate), poly(diallyldimethylammonium chloride),[29] vitamin C,[30] diblock copolymers,[31] and gelatin.[32] In most of the methods including above organic agents, the surfactants/capping agents played a vital role by preferentially adsorbing to some crystal planes of the growing nuclei. Thus, the growth kinetics was ultimately altered; crystal faces with relative stability were grown by promoting or inhibiting crystal growth in some particular planes leading to formation of anisotropic ZnO nanostructures. However, in these studies, controllability over morphology of ZnO was more or less satisfactory.

Surfactants of Triton X series with varying (ethyleneglycoether)$_x$ groups are non-ionic surfactants and are documented for their versatile properties including wetting, detergency, hard metal cleaning, non-toxicity, and excellent emulsification. Triton X-405, a member of Triton X series with $x = 35_{(avg)}$, is an excellent emulsion stabilizer, effective at high temperatures and having good solubility in the presence of salts, electrolytes, and caustic solutions.

Here, in this chapter, we are reporting Triton X-405-mediated facile sonochemical method for synthesis of a newly structured ZnO hierarchical microarchitecture. It is composed of number of nanosheets with a thickness of 10 nm arranged alternately. These ZnO structures showed a high surface to volume ratio and possess stability against aggregation. The as-synthesized ZnO shows enhanced photocatalytic performance which is understood to be induced due to its hierarchical morphology.

3.2 MATERIALS AND METHODS

3.2.1 SYNTHESIS OF HIERARCHICAL ZNO MICROARCHITECTURE

All chemicals used in this study were of analytical grade and were used without further purification. In a typical sonochemical reaction for synthesis, hierarchical ZnO, 1.5 g of zinc (II) acetate dihydrate ($C_4H_6O_4Zn\cdot2H_2O$; Aldrich, 98%), and 2.1 g of citric acid ($C_6H_8O_7$; Aldrich, 98%) were added to 100 mL of deionized water containing premixed (5%) Triton X-405. The solution was then treated with an ultrasonic irradiation operated at dissipated power of 100 W for 1 h (with 5 s ON and 5 s OFF pulse). During ultrasound irradiation, the 20 mL 0.1 M aqueous NaOH solution was added dropwise to the solution within the first 30 min. After 1 h, the white colloidal solution was centrifuged, washed several times with ethanol (chemicals), and then dried in oven at 100°C for 12 h.

3.2.2 CHARACTERIZATION

The as-synthesized product was further characterized by various techniques. Rigaku Miniflex X-ray diffractometer with Cu-Kα irradiation at $\lambda = 1.5406$ Å, a current of 40 mA, and a voltage of 40 kV was used for the X-ray diffraction (XRD) analysis. The morphology was determined by field emission scanning electron microscope (FESEM) (Model JEOL-JSM Model 6700F) attached with energy dispersive spectroscopy (EDS). The room-temperature photoluminescence (PL) spectra of aqueous suspension were recorded on an F-2500 fluorescence spectrophotometer. QUADRASORB "SI" analyzer was used for the N_2 adsorption/desorption isotherms. The JASCO V-570 spectrophotometer was used for measurement of UV–visible absorption spectra of aqueous suspension. The spectral changes in concentration of rhodamine B (RhB) dye during photocatalytic degradation were also studied using the same spectrophotometer.

3.2.3 PHOTOCATALYTIC ACTIVITY MEASUREMENTS

A 100 mL cylindrical pyrex beaker was used as the photoreactor vessel. Prior to sunlight irradiation, the RhB (5.0×10^{-5} M, 50 mL) solution containing the as-synthesized ZnO catalyst (20 mg) was magnetically stirred in the dark in order to achieve the adsorption–desorption equilibrium. The pH value of the RhB aqueous solution (5.0×10^{-5} M) was measured to be about 6.4. The fixed amount of samples were collected from the photoreactor at regulated intervals and then centrifuged at 4000 rpm for 10 min. The centrifuged solution was analyzed in a quartz cell for optical absorbance measurement at absorption maxima of RhB (λ_{max} = 553 nm).

3.3 RESULTS AND DISCUSSIONS

3.3.1 MORPHOLOGY AND STRUCTURE

Typical XRD pattern of as-synthesized ZnO product is depicted in Figure 3.1. From figure, all the peaks match well with the bulk ZnO, and can be indexed to the hexagonal wurtzite structure of ZnO (JCPDF No 36-1451). Further, no crystalline phases related to hydroxide can be observed from the XRD spectrum of the product.

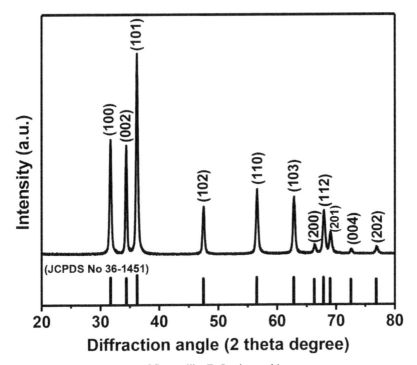

FIGURE 3.1 XRD spectrum of flower-like ZnO microarchitecture.

Figure 3.2 reveals the corresponding morphology of the as-synthesized ZnO. From figure, it can be observed that, the product consists of a large number of microsized flower-like morphology (Fig. 3.2a). A high-magnification FESEM image shows that these flower-like architectures are built-up of large number of interconnected nanosheets in a regulated fashion creating a net-like morphology (Fig. 3.2b). These regularly grown ZnO nanosheets are having thickness of about 20–30 nm and a planar size of about 100 nm. Vigilant observation of these nanosheet-built networks reveals that the nanosheets are connected in a fashion so that there is nearly a 60° angle between them (Fig. 3.2c). The corresponding EDS spectrum of the as-synthesized flower-like ZnO is shown in Figure 3.2d. The as-synthesized product consists of Zn and O only. No peaks related to impurity can be observed from the EDS spectrum indicating the formation of pure ZnO product.

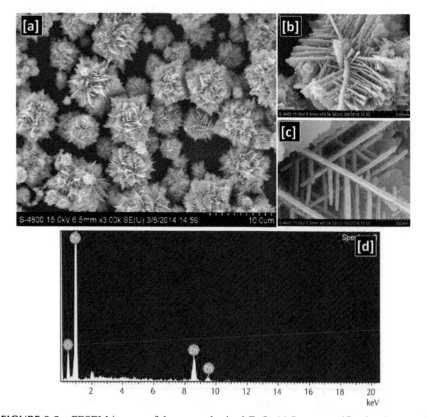

FIGURE 3.2   FESEM images of the as-synthesized ZnO. (a) Low magnification image of the products; (b) the local magnification image of (a); (c) high magnification top-view (b); and (d) corresponding EDS spectrum.

The information about the specific surface area and the pore sizes of the as-synthesized ZnO product was collected using the N_2 adsorption isotherm analysis. From Figure 3.3, the adsorption–desorption isotherm discloses type IV nature, indicating that monolayer formation is missing and there exists abundant mesopores in the architectures.[33] Further, the corresponding pore size distribution curve (shown in inset of Fig. 3.3) reveals the non-uniform pore size distribution with most of the pores ranging from 10 to 30 nm. The specific surface area of the as-synthesized ZnO evaluated by BET equation is found to be ~29.49 m^2g^{-1}.

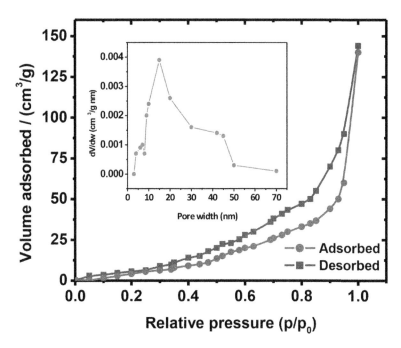

FIGURE 3.3 Nitrogen adsorption–desorption isotherms of as-synthesized flower-like ZnO microarchitecture (inset shows the corresponding pore size distribution curve).

UV–Vis absorption spectra of the as-synthesized flower-like ZnO crystals are shown in Figure 3.4a. From figure, the strong absorption peak centered at 372 nm (3.34 eV) can be observed which is very close to that of the bulk. To investigate the optical properties of the synthesized ZnO further, room-temperature PL studies were carried out. Generally, two peaks in the UV and in visible regions are observed in the PL spectrum of the ZnO. The UV emission originates due to the free-exciton recombination, also called as near band edge emission (NBE), while the green emission, known as deep level emission, appears because of the impurities and structural defects. Room-temperature PL spectrum of as-synthesized ZnO product is displayed in Figure 3.4b.

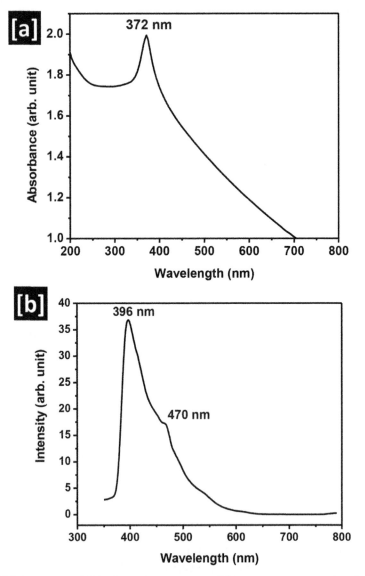

FIGURE 3.4   (a) UV–visible absorption spectrum and (b) room-temperature photoluminescence (PL) spectrum of flower-like ZnO microarchitecture.

From the figure, it can be seen that the PL spectrum consists of a single sharp UV emission at 396 nm. Additionally, weak emission centered at 470 nm located in the blue region could also be observed, which may be due to the transition between the oxygen vacancy and the interstitial oxygen.[33] It is believed that the singly ionized oxygen vacancies in ZnO lead to the green emission and result from the recombination of a photon-generated hole.[34] Ying et al. have reported that the reduced concentration of the oxygen vacancies causes the emergence of NBE with a sharp and strong intensity and a green emission with short or suppressed intensity.[34] In the case of ZnO flower-like architectures demonstrated here also, a strong and sharp NBE emission is observed which indicates that the as-synthesized product possesses less structural defects and good optical properties.

3.3.2 PHOTOCATALYTIC ACTIVITY

Several studies have demonstrated that ZnO has been exploited as a semiconductor photocatalyst for the photocatalytic degradation of water pollutants. Perceptibly, the as-synthesized flower-like ZnO micro/nanoarchitecture is supposed to show a superior photocatalytic activity because of its unusual morphology exhibiting a high surface area (~29.49 m^2g^{-1}). This structure-promoted enhancement of catalytic activity of ZnO has been established by studying photocatalytic degradation of RhB dye aqueous solutions (5.0×10^{-5} M, 50 mL) with 20 mg of the as-synthesized ZnO under sunlight irradiation. When ZnO structures are illuminated by sunlight with energy equal to or greater than the band gap energy; conduction band electrons (e_{cb}^{-}) and valence band holes (h_{vb}^{+}) are generated on the surfaces of the ZnO crystals. At the surface of the ZnO nanostructures, holes can react with water to form highly reactive hydroxyl radicals (OH•), whereas superoxide radical anion (O_2•) can be formed by oxygen by accepting electrons. These superoxide radical anions are also formed. These hydroxyl radicals with powerful oxidation ability can promote the degradation of the dye molecules.[35] Figure 3.5 demonstrates the changes in the absorbance spectra of RhB dye aqueous solutions during photocatalytic degradation under sunlight irradiation in the presence of as-synthesized ZnO product. From figure, the peak corresponding to λ_{max} of RhB molecules appearing at 553 nm decreases rapidly with the exposure time, and completely vanishes after about 40 min signifying the total decay of RhB dye molecules. A sequence of color changes corresponding to the successive absorption measurements during photodegradation is shown in the inset of Figure 3.5. It is clear that the strong magenta color of the initial solution gradually fades with increasing exposure time to sunlight. This superior photocatalytic performance of the hierarchical ZnO can be ascribed to their hierarchical morphology with special structural features and a greater specific surface area.

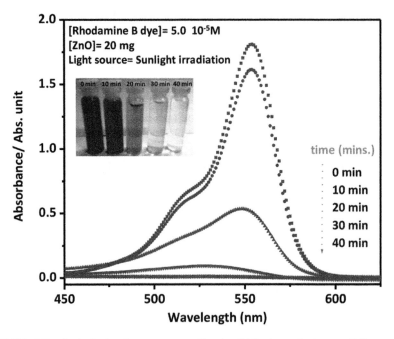

FIGURE 3.5 Optical absorbance spectra for the RhB dye solution at different time intervals of photodegradation (initial concentration: 5.0×10^{-5} M, 50 mL) in the presence of as-synthesized ZnO (20 mg under sunlight irradiation for different durations (inset shows the corresponding time-dependent color change).

3.4 CONCLUSION

In summary, 3D flower-like ZnO composed of nanosheets has been successfully synthesized by a facile sonochemical method. Notably, these flower-like ZnO has displayed a superior photocatalytic performance in the photodegradation of RhB. This is mainly ascribed to their high surface area, and unique 3D morphological features. Further, these flower-like ZnO can be employed as a capable photocatalyst for practical application in the photocatalytic degradation of organic dyes.

KEYWORDS

- **zinc oxide**
- **Triton X**
- **hierarchical morphology**
- **photocatalysis**

REFERENCES

1. Rao, P. M.; Zheng, X. Rapid Catalyst-Free Flame Synthesis of Dense, Aligned α-Fe$_2$O$_3$ Nanoflaks and CuO Nanoneedle Arrays. *Nano Lett.* **2009,** *9,* 3001–3006.
2. Lee, J. C.; Kim, T. G.; Lee, W.; Han, S. H.; Sung, Y. M. Growth of CdS Nanorod-Coated TiO$_2$ Nanowires on Conductive Glass for Photovoltaic Applications. *Cryst. Growth Des.* **2009,** *9,* 4519–4523.
3. Bohle, D. S.; Spina, C. J. Cationic and Anionic Surface Binding Sites on Nanocrystalline Zinc Oxide: Surface Influence on Photoluminescence and Photocatalysis. *J. Am. Chem. Soc.* **2009,** *131,* 4397–4404.
4. Zhang, Y.; Carlos, G. N.; Utke, I.; Michler, J.; Rossell, M. D.; Erni, R. Understanding and Controlling Nucleation and Growth of TiO$_2$ Deposited on Multiwalled Carbon Nanotubes by Atomic Layer Deposition. *J. Phys. Chem. C.* **2015,** *119,* 3379–3387.
5. Gao, X. F.; Jiang, L. Biophysics: Water-Repellent Legs of Water Striders. *Nature.* **2004,** *432,* 36.
6. Xu, X.; Wu, M.; Asoro, M.; Ferreira, P. J.; Fan, D. L. One-Step Hydrothermal Synthesis of Comb-Like ZnO Nanostructures. *Cryst. Growth Des.* **2012,** *12,* 4829–4833.
7. Look, D. C. Recent Advances in ZnO Materials and Devices. *Mater. Sci. Eng. B.* **2001,** *80,* 383–387.
8. Feng, X. J.; Feng, L.; Jin, M.; Zhai, J.; Jiang, L.; Zhu, D. B. Reversible Super-Hydrophobicity to Super-Hydrophilicity Transition of Aligned ZnO Nanorod Films. *J. Am. Chem. Soc.* **2004,** *126,* 62–63.
9. Arnold, M. S.; Avouris, P.; Pan, Z. W.; Wang, Z. L. Field-Effect Transistors Based on Single Semiconducting Oxide Nanobelts. *J. Phys. Chem. B.* **2003,** *107,* 659–663.
10. Kuo, C. L.; Kuo, T. J.; Huang, M. H. Hydrothermal Synthesis of ZnO Microspheres and Hexagonal Microrods with Sheetlike and Platelike Nanostructures. *J. Phys. Chem. B.* **2005,** *109,* 20115–20121.
11. Hoffman, A. J.; Carraway, E. R.; Hoffmann, M. R. Photocatalytic Production of H$_2$O$_2$ and Organic Peroxides on Quantum-Sized Semiconductor Colloids. *Environ. Sci. Technol.* **1994,** *28,* 776–785.
12. Carraway, E. R.; Hoffman, A. J.; Hoffmann, M. R. Photocatalytic Oxidation of Organic Acids on Quantum-Sized Semiconductor Colloids. *Environ. Sci. Technol.* **1994,** *28,* 786–793.
13. Sato, K.; Aoki, M.; Noyori, R. A "Green" Route to Adipic Acid: Direct Oxidation of Cyclohexenes with 30 Percent Hydrogen Peroxide. *Science.* **1998,** *281,* 1646–1647.

14. Ye, C. H.; Bando, Y.; Shen, G. Z.; Golberg, D. Thickness-Dependent Photocatalytic Performance of ZnO Nanoplatelets. *J. Phys. Chem. B.* **2006,** *110,* 15146–15151.

15. Curri, M. L.; Comparelli, R.; Cozzoli, P. D.; Mascolo, G.; Agostiano, A. Colloidal Oxide Nanoparticles for the Photocatalytic Degradation of Organic Dye. *Mater. Sci. Eng. C.* **2003,** *23,* 285–289.

16. Hariharan, C. Photocatalytic Degradation of Organic Contaminants in Water by ZnO Nanoparticles: Revisited. *Appl. Catal. A.* **2006,** *304,* 55–61.

17. Yang, H. G.; Sun, C. H.; Qiao, S. Z.; Zou, J.; Liu, G.; Smith, S. C.; Cheng, H. M.; Lu, G. Q. Anatase TiO$_2$ Single Crystals with a Large Percentage of Reactive Facets. *Nature.* **2008,** *453,* 638.

18. Ohno, T.; Sarukawa, K.; Matsumura, M. Crystal Faces of Rutile and Anatase TiO$_2$ Particles and Their Roles in Photocatalytic Reactions. *New J. Chem.* **2002,** *26,* 1167–1170.

19. Jang, E. S.; Won, J. H.; Hwang, S. J.; Choy, J. H. Fine Tuning of the Face Orientation of ZnO Crystals to Optimize Their Photocatalytic Activity. *Adv. Mater.* **2006,** *18,* 3309.

20. Wang, Z. L.; Kong, X. Y.; Zuo, J. M. Induced Growth of Asymmetric Nanocantilever Arrays on Polar Surfaces. *Phys. Rev. Lett.* **2003,** *91,* 185502–185504.

21. Ozgur, U.; Alivov, Y. I.; Liu, C.; Teke, A.; Reshchikov, M. A.; Dogan, S.; Avrutin, V.; Cho, S. J.; Morkoc, H. J. A Comprehensive Review of ZnO Materials and Devices. *Appl. Phys.* **2005,** *98,* 041301–041301-103.

22. Xu, L. F.; Guo, Y.; Liao, Q.; Zhang, J. P.; Xu, D. S. Morphological Control of ZnO Nanostructures by Electrodeposition. *J. Phys. Chem. B.* **2005,** *109,* 13519–13522.

23. Tian, Z. R.; Voigt, J. A.; Liu, J.; McKenzie, B.; McDermott, M. J.; Rodriguez, M. A.; Konishi, H.; Xu, H. F. Complex and Oriented ZnO Nanostructures. *Nature Mater.* **2003,** *2,* 821–826.

24. Kuo, C. L.; Kuo, T. J.; Huang, M. H. Hydrothermal Synthesis of ZnO Microspheres and Hexagonal Microrods with Sheetlike and Platelike Nanostructures. *J. Phys. Chem. B.* **2005,** *109,* 20115–20121.

25. Cho, S.; Jung, S. H.; Lee, K. H. Morphology-Controlled Growth of Zno Nanostructures Using Microwave Irradiation: From Basic to Complex Structures. *J. Phys. Chem. C.* **2008,** *112,* 12769–12776.

26. Cho, S.; Jang, J. W.; Jung, S. H.; Lee, B. R.; Oh, E.; Lee, K. H. Precursor Effects of Citric Acid and Citrates on ZnO Crystal Formation. *Langmuir.* **2009,** *25,* 3825–3831.

27. Gerstel, P.; Hoffmann, R. C.; Lipowsky, P.; Jeurgens, L. P. H.; Bill, J.; Aldinger, F. Mineralization from Aqueous Solutions of Zinc Salts Directed by Amino Acids and Peptides. *Chem. Mater.* **2006,** *18,* 179–186.

28. Peng, Y.; Xu, A. W.; Deng, B.; Antonietti, M.; Colfen, H. Polymer-Controlled Crystallization of Zinc Oxide Hexagonal Nanorings and Disks. *J. Phys. Chem. B.* **2006,** *110,* 2988–2993.

29. Garcia, S. P.; Semancik, S. Controlling the Morphology of Zinc Oxide Nanorods Crystallized from Aqueous Solutions: The Effect of Crystal Growth Modifiers on Aspect Ratio. *Chem. Mater.* **2007,** *19,* 4016–4022.

30. Cho, S.; Jeong, H.; Park, D. H.; Jung, S. H.; Kim, H. J.; Lee, K. H. The Effects of Vitamin C on ZnO Crystal Formation. *Cryst. Eng. Comm.* **2010,** *12,* 968–976.

31. Munoz-Espí, R.; Jeschke, G.; Lieberwirth, I.; Gomez, C. M.; Wegner, G. Zno-Latex Hybrids Obtained by Polymer-Controlled Crystallization: A Spectroscopic Investigation. *J. Phys. Chem. B.* **2007,** *111,* 697–707.

32. Bauermann, L. P.; del Campo, A.; Bill, J.; Aldinger, F. Heterogeneous Nucleation of ZnO Using Gelatin as the Organic Matrix. *Chem. Mater.* **2006,** *18,* 2016–2020.

33. Lu, H.; Wang, S.; Zhao, L.; Li, J.; Dong, B.; Xu, Z. Hierarchical ZnO Microarchitectures Assembled by Ultrathin Nanosheets: Hydrothermal Synthesis and Enhanced Photocatalytic Activity. *J. Mater. Chem.* **2011**, *21*, 4228–4234.
34. Ying, D.; Zhang, Y.; Bai, Y. Q.; Wang, Z. L. Bicrystalline Zinc Oxide Nanowires. *Chem. Phys. Lett.* **2003**, *375*, 96–101.
35. Hoffmann, M. R.; Martin, S. T.; Choi, W.; Bahnemann, D. W. Environmental Applications of Semiconductor Photocatalysis. *Chem. Rev.* **1995**, *95*, 69–96.

CHAPTER 4

SONOCHEMICAL SYNTHESIS OF Mg-DOPED ZnO NPS FOR EFFICIENT SUNLIGHT DRIVEN PHOTOCATALYSIS

S. P. MESHRAM[1], P. D. JOLHE[2,*], S. D. SHINGTE[3], B. A. BHANVASE[4], and S. H. SONAWANE[5]

[1]Department of Nanoscience and Nanotechnology, Centre for Materials for Electronics Technology, Pune 411008, Maharashtra, India

[2]Department of Biotechnology, Sinhgad College of Engineering, Savitribai Phule Pune University, Pune 411041, Maharashtra, India

[3]Department of Chemical Engineering, Faculty of Applied Sciences, TU Delft, Delft, The Netherlands

[4]Chemical Engineering Department, Laxminarayan Institute of Technology, Rashtrasant Tukadoji Maharaj Nagpur University, Nagpur 440033, Maharashtra, India

[5]Chemical Engineering Department, National Institute of Technology, Warangal 506004, Telangana, India

*Corresponding author. E-mail: pdjolhe.scoe@sinhgad.edu

CONTENTS

ABSTRACT

This chapter highlights succinic acid mediated facile sonochemical method for synthesis of Mg-doped ZnO nanoparticles. The incorporation of Mg into ZnO matrix was investigated using X-ray diffraction (XRD) technique. Textural properties of as-synthesized undoped and Mg-doped ZnO samples were investigated using FESEM coupled with elemental mapping. Optical property investigations revealed that as the Mg-doping content increased up to 0.09%, the absorption edge slightly shifted to shorter wavelengths. Further increase in Mg-doping concentration (i.e., for 0.12%) experienced a shift toward higher wavelength. Additionally, reduced near band edge emission (NBE) and excitonic emission intensities were observed for the most active photocatalyst compared to undoped ZnO. Mg-doped ZnO (0.09 *at*%) nanoparticles (NPs) exhibited highest photocatalytic activity among all the synthesized samples and can degrade the methylene blue (MB) dye solution in merely 60 min under sunlight irradiation.

4.1   INTRODUCTION

The growing population worldwide has resulted in increased water pollution. The ever-growing problems related to environment have led to intensive research in the area of environmental remediation.[1–3] Recently, considerable advances have been made in the area of nanostructured semiconductor-mediated photocatalytic applications.[4–7] Among several semiconductor metal oxides used in advanced oxidation processes, ZnO is reflected as a benchmark catalyst due to its non-toxic nature, excellent physical–chemical stability, and low cost.[8] ZnO can be operated over a large fraction of solar spectrum compared to TiO_2.[9]

In general, several factors such as morphology, nature of dopants, phase purity, crystallite size, surface area, and preparation method are accountable for photocatalytic activity of ZnO. Despite of high photocatalytic activity of ZnO, the narrow band gap associated with it declines its use as photocatalyst under sunlight irradiation.

Several modification methods have been developed to improve the photocatalytic activity of ZnO. Band gap engineering is one of the most important approaches involved in designing optically and electrically confined structures. Several researchers have attempted a transition metal and anionic doping to improve the photocatalytic activity of ZnO.[10–12] Still, surplus charge carrier recombination limits the practical applications of these transition metal- and anion-doped ZnO. Optionally, II–VI compounds like magnesium oxide (MgO) having high optical band gap can be used for alloying ZnO to increase its band gap, owing to small lattice mismatch of MgO with ZnO. It has been reported that, the band gap can be considerably tuned from 3.3 to 7.8 eV for wurtzite and cubic structured $Zn_{1-x}Mg_xO$, by simply varying the Mg content.[13] Although, solubility limits of Mg in ZnO is low (about ~4 *at*%), it largely depend on the synthesis techniques and the processing conditions.[14] Accordingly, several researchers have reported different techniques to intricate $Zn_{1-x}Mg_xO$, namely sol–gel method, chemical vapor deposition, pulsed laser deposition, spray pyrolysis, molecular beam epitaxy (MBE), radio frequency (RF) magnetron sputtering, electrodeposition, and metal-organic vapor-phase epitaxy (MOVPE).[15–22]

Morphology and crystallite size are the key parameters of nanomaterials and generally can be tuned by simply varying synthesis methods. Although a variety of methods have been used to synthesize $Zn_{1-x}Mg_xO$ nanomaterials, a great deal of work is still needed to explore low cost and new approach. Recently, numerous works have appeared in the literature describing the sonochemical synthesis of a variety of inorganic materials because of its green and practical uniqueness. In this report, we utilized sonochemical method to synthesize the band gap modified Mg-doped ZnO nanoparticles (NPs). The detailed investigation on the structural, optical, and photocatalytic properties of the band gap-modified Mg-doped ZnO NPs is also carried.

4.2 MATERIALS AND METHODS

4.2.1 SYNTHESIS OF $Zn_{1-x}Mg_xO$ NPs

Facile sonochemical method was used to synthesize undoped and doped $Zn_{1-x}Mg_xO$ (x = 0–0.12) NPs. All the reagents used for synthesis were of analytical grade. Zinc nitrate $Zn(NO_3)_2.6H_2O$, $Mg(NO_3)_2.6H_2O$, and succinic acid $((CH_2)_2(CO_2H)_2)$ were firstly dissolved in distilled water. During synthesis, the molar ratio of the metal nitrate and succinic acid was maintained at 1:1.56. After dissolving the precursors in the solvent, the solution was treated under ultrasonic irradiation (Fisher Scientific) at 45 W of dissipated power for 1 h (with 5 s ON and 5 s OFF pulse). During ultrasound irradiation, the 20 mL, 0.1 M aq. NaOH solution was added dropwise to the solution within the first 30 min. After 1 h, the white colloidal solution so obtained was centrifuged, washed several times with water and ethanol to remove impurities and then dried in oven at 100°C for 12 h.

4.2.2 *CHARACTERIZATION TECHNIQUES*

The as-synthesized products were further characterized by various techniques. Rigaku Miniflex X-ray diffractometer with Cu-Kα irradiation at $\lambda = 1.5406$ Å, a current of 40 mA, and a voltage of 40 kV was used for the X-ray powder diffraction (XRD) analysis. The morphology was determined by field emission scanning electron microscope (FESEM) (Model JEOL-JSM Model 6700F) attached with energy dispersive spectroscopy (EDS). The room-temperature photoluminescence (PL) spectra of aqueous suspension were recorded on an F-2500 Fluorescence spectrophotometer. The JASCO V-570 spectrophotometer was used for measurement of UV–visible absorption spectra of aqueous suspension. The spectral changes in concentration of methylene blue (MB) dye during photocatalytic degradation were also studied using the same spectrophotometer.

4.2.3 *PHOTOCATALYTIC ACTIVITY MEASUREMENTS*

The photocatalytic activities of as-synthesized NPs were assessed by the sunlight-mediated photodegradation of MB dye in aqueous solution. In a typical experiment, 0.2 g of catalyst sample was dispersed into 100 mL (100 ppm) MB solution. Prior to sunlight irradiation, the suspended solution was stirred for 30 min in the dark to achieve the adsorption equilibrium of MB on the catalysts. During photocatalytic degradation experiments, the samples were taken out from the mixed suspension at regulated intervals of 5 min. The samples were then centrifuged to remove the catalysts at 5000 rpm for 10 min. The photodegradation efficiency was monitored by assessing the absorbance with UV–Vis spectrophotometer at its maximum absorption wavelength of 670 nm.

4.3 RESULTS AND DISCUSSION

4.3.1 *STRUCTURAL AND MORPHOLOGY ANALYSES*

XRD patterns of $Zn_{1-x}Mg_xO$ ($x = 0$, 0.03, 0.06, 0.09, and 0.12) NPs synthesized by facile sonochemical method are depicted in Figure 4.1. From figure, it can be seen that all the synthesized NPs exhibit hexagonal wurtzite structure. The diffraction peaks appearing at about $2\theta = 31.8°$, $34.5°$, and $36.3°$ can be assigned to (1 0 0), (0 0 2), and (1 0 1) peaks of ZnO, respectively.[23,24]

All the diffraction peaks in the patterns matches well to würtzite ZnO, with good crystallite quality and having a preferred orientation in (1 0 1) plane. Further, there are not any peaks associated to Mg phases which indicate the incorporation of Mg ions into the ZnO lattice. This can be attributed to quite close ionic radius of Mg^{2+} (0.057 nm) to that of Zn^{2+} (0.060 nm), which makes the incorporation of Mg^{2+} ions in the ZnO lattice possible. The Mg-doped ZnO NPs synthesized by sonochemical method are found to be a single-phase alloy over a wide range of Mg-doping levels.

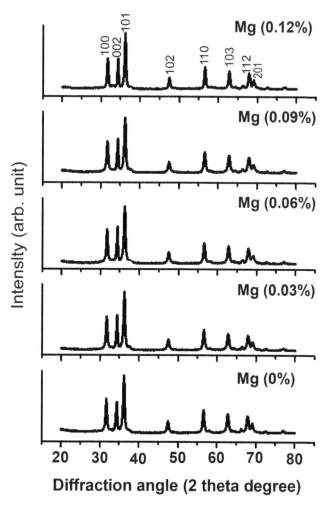

FIGURE 4.1   X-ray diffraction patterns of $ZnO_{1-x}Mg_xO$ (x = 0, 0.03, 0.06, 0.09, and 0.12) NPs.

The surface morphologies of as-synthesized $Zn_{1-x}Mg_xO$ NPs were further characterized by FESEM and the results are displayed in Figure 4.2. The FESEM images (Fig. 4.2a–j) clearly show that Mg doping slightly altered the sizes of ZnO NPs. The Mg-doped ZnO NPs with different Mg content (0.03–0.12) (Fig. 4.2c–j) exhibits uniform sizes. Mg-doped ZnO NPs; exhibits discrete vesicular morphology resulted by the ultrasound

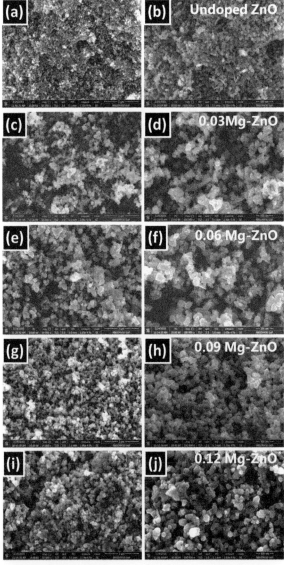

FIGURE 4.2   FESEM images of undoped (a–b) and doped ZnO (c–j) NPs.

irradiation during sonochemical synthesis. The average particle size (around 50–60 nm) and uniform morphology were observed for Mg-doped ZnO NPs with different Mg content (0.03, 0.06, 0.09, and 0.12). Although, the solubility limit of Mg in ZnO is reported to be as low as 4 *at*%, the combined XRD and FESEM results, demonstrated the ability of sonochemical method to effectively enhance the solubility limit of MgO in ZnO.

The Mg-doped ZnO NPs were further characterized by EDAX, in order to confirm the composition of doped sample. Figure 4.3 displays the EDAX pattern and elemental mapping of representative 0.09% Mg-doped ZnO sample. From the figure, it can be observed that, the sample is composed of Mg, Zn, and O only. Also, the calculated percentage of Mg is in accordance with the quantities used in the synthetic architecture for preparation of $Zn_{99.91}Mg_{0.09}O$ sample.

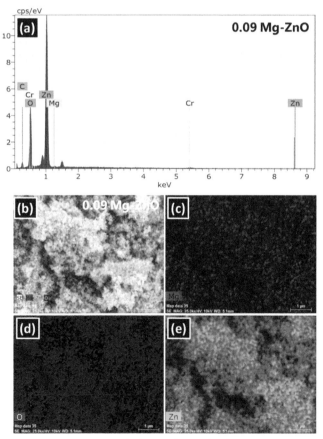

FIGURE 4.3 EDAX spectrum (a) and elemental mapping images (c–e) of Mg-doped ZnO sample (Mg = 0.09).

4.3.2   OPTICAL PROPERTIES OF MG-DOPED ZNO NPS

The as-synthesized Mg-doped ZnO NP samples were characterized by UV–visible absorption spectroscopy to determine the band gap. The optical absorption spectra analysis of ZnO and $Zn_{1-x}Mg_xO$ NPs were done by dispersing the samples in isopropyl alcohol. Figure 4.4 displays the UV–visible absorption spectra of all the $Zn_{1-x}Mg_xO$ ($x = 0$–0.12) samples. From figure, a clear absorption edge in the UV–Vis region can be observed for all the samples. It can also be seen that, as the Mg content increased up to 0.09% Mg concentration, the absorption edge slightly shifted to short wavelengths. Further increase in Mg concentration (i.e., for 0.12%) leads

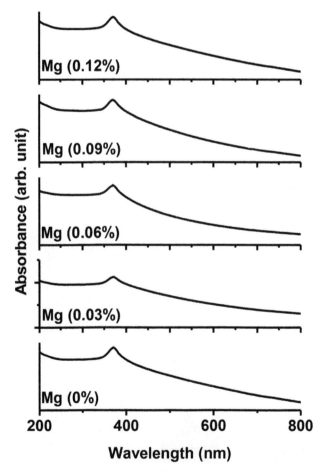

FIGURE 4.4   UV–Vis absorbance spectra of $Zn_{1-x}Mg_xO$ ($x = 0$–0.12) NPs.

to shift the absorption edge to higher wavelength. Absorption spectra of solids are contributed by various processes, one of them in which the lattice or electrons absorb the incident photo energy and the transferred energy is conserved. Here in present case, the initial blue shift in the optical absorption spectra may be originated due to Burstein–Moss effect. Up to the 0.09 *at%* Mg-doping concentration, the shifting of Fermi level into the conduction band will be caused due to increase of carrier concentration. This further resulted in a blue shift in the near band edge (NBE) emission. Since, the size effect has trivial influence on the band structure of ZnO NPs (for diameter > 7 nm), the observed red shift for the higher Mg-doping concentration (> 0.09 *at%*) may be attributed to the concentration of doping atoms.[25] At higher Mg concentration (i.e., above 0.09 *at%*), more Mg^{2+} ions will be available for replacing the Zn^{2+} ions, thereby the strong *sp–d* exchange interaction between free charge carriers and the localized magnetic moments will be facilitated in the semiconductor band. This will further lead to red shift of the absorption edge. The optical band gap energies were determined using the equation: Band gap (eV) = $1240/\lambda_{max}$. The band gap values calculated for undoped and Mg-doped ZnO was found to be 3.30, 3.31, 3.32, 3.34, and 3.33 eV, respectively.

4.3.3 PHOTOCATALYTIC ACTIVITY

The photocatalytic activity of the as-synthesized undoped and Mg-doped ZnO NPs was evaluated by the photocatalytic degradation of MB under sunlight irradiation. Photocatalytic degradation of a 1×10^{-5} M MB solution as a function of sunlight irradiation time on undoped and Mg-doped ZnO NPs is shown in Figure 4.5a. A blank experiment without any catalyst was also performed to ensure that the degradation of MB did not appear from photolysis. A control experiment revealed that under sunlight irradiation, MB does not lead to any perceptible degradation. On the other hand, in the presence of undoped ZnO, the characteristic absorption band of MB decreases with increasing exposure time and the degradation is invariable after 120 min (Fig. 4.5b). Further, the Mg-doped ZnO NPs catalyzed degradation of MB greater than undoped ZnO. Thus, 0.09 *at%* Mg-doped ZnO NPs exhibited highest photocatalytic activity among all the synthesized samples and can degrade the MB dye solution in merely 60 min under sunlight irradiation. Generally, the photocatalytic degradation of a dye in presence of semiconductor oxide is supposed to trail the following mechanism. Upon irradiation of UV–visible light, the ZnO semiconductor absorbs photons with energy equal to or greater than its band gap. Thus, the electrons in the valence band are excited to the conduction band leaving behind a hole in the valence band according to following equations:

$$ZnO \rightarrow e^-_{cb} + h^+_{vb} \tag{4.1}$$

$$e^-_{cb} + O_2 \rightarrow {}^\bullet O_2^- \tag{4.2}$$

$$h^+_{vb} + OH^- \rightarrow {}^\bullet OH \tag{4.3}$$

$${}^\bullet O_2^- + H_2O \rightarrow HO_2^\bullet + OH^- \tag{4.4}$$

$$HO_2^\bullet + H_2O \rightarrow H_2O_2 + {}^\bullet OH \tag{4.5}$$

$$H_2O_2 \rightarrow 2^\bullet OH \tag{4.6}$$

$${}^\bullet OH + organic\ compound \rightarrow CO_2 + H_2O \tag{4.7}$$

where h^+_{vb} and e^-_{cb} are the electron vacancies in the valence band and the electron in the conduction band, respectively.

Thus, during photocatalytic degradation, the ${}^\bullet OH$ radical and ${}^\bullet O_2^-$ group play a significant role. In addition to this, particle sizes,[26] morphologies,[27] surface properties,[28] and bandgap energies[29] also affect the photocatalytic activity of semiconductor oxide. Since, the size and morphology of as-synthesized Mg-doped ZnO samples are nearly indistinguishable (according to FESEM in Fig. 4.2), these surface properties cannot be considered for the progress of photocatalytic activity. Significantly, improved photocatalytic activities of as-synthesized Mg-doped ZnO can be explained on the basis of their band gaps. Formation of wide band gap $Zn_{1-x}Mg_xO$ photocatalysts is clearly evident from UV−Vis spectroscopy results. In general, photoexcited electron−hole pairs with higher redox potential are caused due to increase in the band gap values, which significantly contributes to increased photocatalytic activity.

PL spectra can be utilized to evaluate presence of surface states, efficiency of charge carrier trapping, and the recombination kinetics.[30] Previously, several researchers have reported correlation of electron−hole pairs in semiconductor particle and photocatalytic activity on the basis of PL spectra.[6,7,30] Investigation of the photocatalytic activities of Mg-doped ZnO NPs was also carried on the basis of PL spectroscopy. PL spectra of representative undoped ZnO, the most active photocatalyst 0.09-MgZnO and the least active sample 0.12-MgZnO are presented in Figure 4.5c. From the figure, reduced NBE and excitonic emission intensities for the most active photocatalyst 0.09-MgZnO compared to undoped ZnO can be clearly seen.

This can be ascribed to widening of band gap due to efficient separation of photoinduced charge carriers. Moreover, in case of the least active photocatalyst 0.12-MgZnO, higher intensities were observed, indicating faster recombination of photoinduced electron–hole pairs. Doping ZnO with Mg in variable concentrations leads to significant differences in the PL intensities and further confirmed a change in the electronic structure of ZnO. Thus to summarize, increased band gap values and efficient electron–hole separation are the key factors accountable for the enhanced sunlight-driven photocatalytic activities of Mg-doped ZnO.

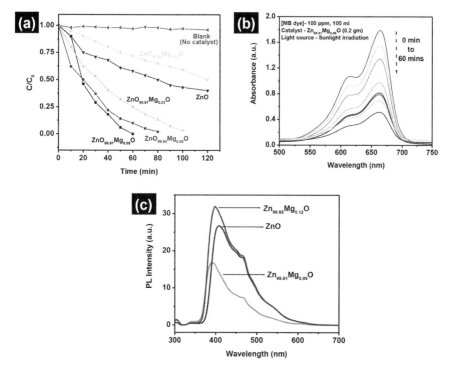

FIGURE 4.5 (a) Comparative photocatalytic activity of ZnO:Mg NPs under sunlight irradiation, (b) absorption spectra changes methylene blue (MB) degradation using $Zn_{99.91}Mg_{0.09}O$ catalyst under sunlight irradiation, and (c) photoluminescence spectra of representative undoped ZnO, $Zn_{99.91}Mg_{0.09}O$, and $Zn_{99.88}Mg_{0.12}O$ samples.

4.4 CONCLUSION

Mg-doped ZnO NPs were synthesized using succinic acid-mediated sonochemical method. XRD results revealed that all the synthesized samples exhibited pure würtzite structure without any Mg-related impurities. The photocatalytic activities of as-synthesized samples were evaluated by the photodegradation of MB under sunlight irradiation. Structural and optical analyses show that the 0.09 $at\%$ Mg doping is optimal for best photocatalytic activity. The facile sonochemical route employed here demonstrated beneficial effect for synthesis of Mg-doped ZnO samples and their utilization as efficient photocatalysts with a low cost for the environmental remediation.

KEYWORDS

- Mg-doped ZnO
- sonochemical
- photoluminiscence
- photocatalysis

REFERENCES

1. Sharma, A.; Rao, P.; Mathur, R. P.; Ameta S. C. Photocatalytic Reactions of Xylidine Ponceau on Semiconducting Zinc Oxide Powder. *J. Photochem. Photobiol. A: Chem.* **1995**, *86*, 197–200.
2. Khodja, A. A.; Sehili, T.; Pilichowski, J. F.; Boule, P. Photocatalytic Degradation of 2-Phenylphenol on TiO$_2$ and ZnO in Aqueous Suspensions. *J. Photochem. Photobiol. A: Chem.* **2001**, *141*, 231–239.
3. Akyol, A.; Yatmaz, H. C.; Bayramoglu, M. Photocatalytic Decolorization of Remazol Red RR in Aqueous ZnO Suspensions. *Appl. Catal. B: Environ.* **2004**, *54*, 19–24.
4. Kamat, P. V. Meeting the Clean Energy Demand: Nanostructure Architectures for Solar Energy Conversion. *J. Phys. Chem. C.* **2007**, *111*, 2834–2860.
5. Mills, A.; Lee, S. K. A Web-Based Overview of Semiconductor Photochemistry-Based Current Commercial Applications. *J. Photochem. Photobiol. A.* **2002**, *152*, 233–247.
6. Etacheri, V.; Seery, M. K.; Hinder, S. J.; Pillai, S. C. Oxygen Rich Titania: A Dopant Free, High Temperature Stable, and Visible-Light Active Anatase Photocatalyst. *Adv. Funct. Mater.* **2011**, *21*, 3744–3752.
7. Etacheri, V.; Seery, M. K.; Hinder, S. J.; Pillai, S. C. Highly Visible Light Active TiO$_{2-x}$N$_x$ Heterojunction Photocatalysts. *Chem. Mater.* **2010**, *22*, 3843–3853.
8. Ismail, A. A.; El-Midany, A. A.; Abel-Aal, E. A.; El-Shall H. Application of Statistical Design Strategies to Optimize the Preparation of ZnO NPs via Hydrothermal Technique. *Mater. Lett.* **2005**, *59*, 1924–1928.
9. Chakrabarti, S.; Dutta, B. K. Photocatalytic Degradation of Model Textile Dyes in Wastewater Using ZnO as Semiconductor Catalyst. *J. Hazard. Mater.* **2004**, *112*, 269–278.

10. Li, G. R.; Bu, Q.; Zheng, F. L.; Su, C. Y.; Tong, Y. X. Electrochemically Controllable Growth and Tunable Optical Properties of $Zn_{1-x}Cd_xO$ Alloy Nanostructures. *Cryst. Growth Des.* **2009,** *9,* 1538–1545.

11. Luo, T.; Yang, Y. C.; Zhu, X. Y.; Chen, G.; Zeng, F.; Pan, F. Enhanced Electromechanical Response of Fe-Doped Zno Films by Modulating the Chemical State and Ionic Size of the Fe- Dopant. *Phys. Rev. B.* **2010,** *82,* 14116–14200.

12. Wang, H. B.; Wang, H.; Zhang, C.; Yang, F. J.; Duan, J. X.; Yang, C. P.; Gu, H. S.; Zhou, M. J.; Li, Q.; Jiang, Y. Preparation and Characterization of Mn and (Mn, Cu) Co-Doped ZnO Nanostructures. *J. Nanosci. Nanotechnol.* **2009,** *5,* 3308–3312.

13. Li, M. G.; Hari-Bala, X. T.; Lv, X. K.; Ma, F.; Sun, L. Q.; Tang Z. C. Direct Synthesis of Monodispersed ZnO Nanoparticles in an Aqueous Solution. *Mater. Lett.* **2007,** *61,* 690–693.

14. Segnit, E. R.; Holland A. E. The System $MgO-ZnO-SiO_2$. *J. Am. Ceram. Soc.* **1965,** *48,* 409–413.

15. Ogata, K.; Koike, K.; Tanite, T.; Komuro, T.; Yan, F.; Sasa, S.; Inoue, M.; Yano, M. ZnO and ZnMgO Growth on *a*-Plane Sapphire by Molecular Beam Epitaxy. *J. Cryst. Growth.* **2003,** *251,* 623–627.

16. Park, W. I.; Yi, G. C.; Jang, H. M. Metalorganic Vapor-Phase Epitaxial Growth and Photoluminescent Properties of Zn1− xMgxO (0⩽ x⩽ 0.49) Thin Films. *Appl. Phys. Lett.* **2001,** *79,* 2022–2024.

17. Wang, P.; Chen, N.; Yin, Z.; Dai, R.; Bai, Y. p-Type $Zn_{1-x}Mg_xO$ Films with Sb Doping by Radio-Frequency Magnetron Sputtering. *Appl. Phys. Lett.* **2006,** *89,* 202102–202111.

18. Sahay, P. P.; Tewari, S.; Nath, R. K. Optical and Electrical Studies on Spray Deposited ZnO Thin Films. *Cryst. Res. Technol.* **2007,** *42,* 723–729.

19. Bellingeri, E.; Marre, D.; Pallecchi, I.; Pellegrino, L.; Canu, G.; Siri, A. S. Deposition of ZnO Thin Films on $SrTiO_3$ Single-Crystal Substrates and Field Effect Experiments. *Thin Solid Films.* **2005,** *486,* 186–190.

20. Lin, C. W.; Cheng, T. Y.; Chang, L.; Juang J. Y. Chemical Vapor Deposition of Zinc Oxide Thin Films on Y_2O_3/Si Substrates. *Phys. Stat. Sol.* **2004,** *1,* 851.

21. Fahoume, M.; Maghfoul, O.; Aggour, M.; Hartiti, B.; Chraibi, F.; Ennaoni, A. Growth and Characterization of ZnO Thin Films Prepared by Electrodeposition Technique. *Sol. Energy Mater. Sol. Cells.* **2006,** *90,* 1437–1444.

22. Srinivasan, G.; Kumar J. Optical and Structural Characterisation of Zinc Oxide Thin Films Prepared by Sol-Gel Process. *Cryst. Res. Technol.* **2006,** *41,* 893–896.

23. Moghaddam, F. M.; Saeisian H. Controlled Microwave-Assisted Synthesis of ZnO Nanopowder and its Catalytic Activity for *O*-Acylation of Alcohol and Phenol. *Mater. Sci. Eng. B.* **2007,** *139,* 265–269.

24. Tabib, A.; Sdiri, N.; Elhouichet, H.; Férid, M. Investigations on Electrical Conductivity and Dielectric Properties of Na Doped ZnO Synthesized from Sol Gel Method. *J. Alloys Compd.* **2015,** *622,* 687–694.

25. Meulenkamp, E. A. Synthesis and Growth of ZnO Nanoparticles. *J. Phys. Chem. B.* **1998,** *102,* 5566–5572.

26. Tomar, M. S; Melgarejo, R.; Dobal, P. S.; Katiyar, R. S. Synthesis of $Zn_{1-x}Mg_xO$ and its Structural Characterization. *J. Mater. Res.* **2001,** *16,* 903–906.

27. Li, D.; Haneda, H. Morphologies of Zinc Oxide Particles and their Effects on Photocatalysis. *Chemosphere.* **2003,** *51,* 129–137.

28. Etacheri, V.; Roshan, R.; Kumar, V. Mg-doped ZnO Nanoparticles for Efficient Sunlight-Driven Photocatalysis. *ACS Appl. Mater. Interfaces.* **2012,** *4,* 2717–2725.

29. Jing, L. Q.; Xu, Z. L.; Sun, X. J.; Shang, J.; Cai, W. M. The Surface Properties and Photocatalytic Activities of ZnO Ultrafine Particles. *Appl. Surf. Sci.* **2001,** *180,* 308–314.
30. Liqiang, J.; Yichun, Q.; Baiqi, W.; Shudan, L.; Baojiang, J.; Libin, Y.; Wei, F.; Honggang, F.; Jiazhong, S. Review of Photoluminescence Performance of Nano-Sized Semiconductor Materials and its Relationships with Photocatalytic Activity. *Sol. Energy Mater. Sol. Cells.* **2006,** *90,* 1773–1787.

CHAPTER 5

INTENSIFIED REMOVAL OF Cu^{2+} AND Fe^{2+} USING GREEN ACTIVATED CARBON DERIVED FROM *LANTANA CAMARA* STEM AND SOYA HULL AND ITS COMPARISON WITH COMMERCIAL ACTIVATED CARBON

A. A. KADU[1], B. A. BHANVASE[1,*], and S. H. SONAWANE[2]

[1]*Department of Chemical Engineering, Laxminarayan Institute of Technology, Rashtrasant Tukadoji Maharaj Nagpur University, Nagpur 440033, Maharashtra, India*

[2]*Department of Chemical Engineering, National Institute of Technology Warangal, Warangal 506004, Telangana, India*

Corresponding author. E-mail: bharatbhanvase@gmail.com

CONTENTS

ABSTRACT

In this study, the preparation of activated carbons from agricultural waste such as soya hull and *Lantana camara* stems were carried out by the chemical activation with H_2SO_4. The prepared adsorbents were analyzed using techniques such as Fourier transform infrared spectroscopy (FTIR) and Brunauer–Emmett–Teller (BET) surface area and used as adsorbents for the removal of Cu^{2+} and Fe^{2+} metal ions from aqueous solution. The effect of adsorbent dose and initial metal ion concentration on the removal efficiency of Cu^{2+} and Fe^{2+} metal ions were studied. The adsorption data was well fitted with the use of Freundlich isotherm models and constant values were estimated for all the cases. The value of Freundlich constant (n) for various cases is ranging from 0 to 1 which is an indication of a favorable condition for multilayer adsorption.

5.1 INTRODUCTION

In the recent years, the increase of heavy metals concentration in water is essentially because of the discharge of wastewater from various industries. Natural water pollution by heavy metal ions has become an important concern all over the world because metal concentrations in water frequently exceed the permissible limit. These heavy metals are hostilely disturbing our environment system because of their toxicological and physiological effects.[1–3] Heavy metals such as arsenic (As), copper (Cu), cadmium (Cd), chromium (Cr), iron (Fe), nickel (Ni), zinc (Zn), lead (Pb), mercury (Hg), and manganese (Mn) are key pollutants of natural water due to their toxic, non-biodegradable, and persistent nature.[4] The industrial sources of these heavy metal ions are mining, smelting, surface finishing, electroplating, electrolysis, electric appliances, electric circuits, fertilizers, and pesticides industries.[5] Out of these heavy metals, Fe and Cu are the important heavy metals that are polluting the environment and also are affecting the human health. Cu^{2+} ions are toxic and carcinogenic when consumed in excess. The excess dose of Cu^{2+} can result in its deposition in liver which leads to various problems like vomiting, nausea, headache, abdominal pain, respiratory problems, liver and kidney failure, and finally, gastrointestinal bleeding.[6] Also, Fe ions can harm tissues by catalyzing the conversion of hydrogen peroxide to free-radical ions that attack cellular membranes, proteins, and DNA.[7] Therefore, there is a dire need of separation of these heavy metals from wastewater.

There are various methods available for the removal of heavy metals from wastewater. These methods are ion-exchange, adsorption, chemical precipitation, coagulation–flocculation, membrane filtration, flotation, and electro-chemical methods.[8] Conversely, the limitations of many of these technologies are high investment costs, high maintenance costs, secondary pollution (generation of toxic sludge, etc.), and complicated process for the treatment of heavy metals.[9] Out of these methods, adsorption of heavy metals on agricultural waste adsorbent is a cost effective and simple method. Biosorption is the process of adsorption of heavy metals on biological materials and it is an important method which has the potential for toxic metal removal from aqueous system. Instead of using biomass as it is for adsorption, it is more effective to use activated carbon generated from such type of biomass.[10,11] Further, because of stringent environmental laws, nowadays, decontamination of wastewater has generated a large interest in the researchers regarding the use of activated carbons prepared from agricultural waste for the adsorptive removal of metal ions. Examples of agricultural wastes that have been effectively used in the preparation of activated carbons are coconut shells, wood chips, saw-dust, corn cobs, coffee husks, and pine tree among others.[12–14]

The main objective of this study was to develop a new activated carbons from *Lantana camara* stem and soya hull by chemical activation with H_2SO_4. The biosorption of heavy metals, that is, Cu^{2+} and Fe^{2+} have been studied with various parameters such as initial concentration, time, and adsorbent dose. Further, the comparison of biosorption capacity of prepared activated carbons in this study was done with commercial activated charcoal. The kinetics and Freundlich isotherm for the sorption of Cu^{2+} and Fe^{2+} on the prepared activated carbon samples were carried out to determine the mechanisms involved in the biosorption process.

5.2 EXPERIMENTAL METHODOLOGY

5.2.1 MATERIALS

The chemicals and reagents used in the experimentation reported in this manuscript were of acid red (AR) grade. Cupric sulfate ($CuSO_4.5H_2O$), ferrous sulfate ($FeSO_4.7H_2O$), and sulfuric acid (H_2SO_4) were purchased from Merck, India and used as received. Commercial activated charcoal (granular LR 2.0–5.0 mm) was purchased from S. D. Fine-Chem Ltd., Mumbai and used as received. The natural low-cost materials, soya hull, was obtained from Ganesh Oil Mill, Dhamangaon, Amravati, Maharashtra, India. *L. camara* stems were obtained from farm, Amravati, Maharashtra, India. The solutions of required concentrations were prepared by suitable dilution of the stock solutions. After desired adsorption, the supernatant solutions were filtered with Whatman filter paper number 1. The concentration of heavy metal was measured with the help of UV–Vis spectrometer (UV-1800 UV–VIS spectrophotometer).

5.2.2   PREPARATION OF ACTIVATED CARBON FROM SOYA HULL AND LANTANA CAMARA

The obtained soya hulls were extensively washed with water to remove oil and dust particles. It was then sundried, and crushed into smaller size. It was further treated with 0.5 N H_2SO_4 for 12 h at 30°C in order to activate it. The treated soya hull was washed with water repeatedly till pH reaches to seven and then dried in oven for 8 h at 100°C. The dried soya hull was then carbonized using an electric furnace at 290°C for 30 min. The obtained activated carbon was grinded and sieved to 200 mesh size. It was later stored inside a desiccator to avoid moisture absorption. Similar procedure was followed for the preparation of activated carbon for *L. camara* stems.

5.2.3   BATCH ADSORPTION EXPERIMENTS

The adsorption of heavy metals namely Cu^{2+} and Fe^{2+} on three different adsorbents, that is, activated charcoal, soya hull activated carbon, and *L. camara* activated carbon was studied by batch adsorption technique. A known quantity of prepared adsorbent (e.g., 0.6 g adsorbent) was added to 100 mL of the heavy metal (namely Cu^{2+} or Fe^{2+}) solution. The effects of initial concentration (10, 20, and 50 ppm) and adsorbent dose (0.2–0.6 g) on percentage removal of Cu^{2+} and Fe^{2+} were investigated at a fixed temperature (30°C). The time of contact was kept constant at 30 min in each batch experiment. The samples were collected at the end of 30 min of contact time and the adsorbent suspension was filtered using Whatman No. 1 filter paper in order to separate the adsorbent. The heavy metal ions concentration remaining in solution was measured with the help of UV–Vis spectrometer (UV-1800 UV–Vis spectrophotometer). The percentage of Cu^{2+} and Fe^{2+} removal and equilibrium adsorption capacity (q_e) were calculated using following equations:

$$\% \text{ removal} = \frac{C_o - C_e}{C_o} \quad (5.1)$$

$$q_e = \frac{V(C_o - C_e)}{m} \quad (5.2)$$

where q_e (mg/g) is the equilibrium adsorption capacity, C_e (mg/L) is the ion concentration at equilibrium, V (L) is the volume of solution, and m (g) is the weight of adsorbent.

5.2.4 MATERIAL CHARACTERIZATIONS

The Fourier transform infrared spectroscopy (FTIR) analysis of activated charcoal, soya hull activated carbon, and *L. camara* activated carbon was performed on FTIR spectroscopy (Shimadzu) in order to investigate the functional groups present on its surface. The specific surface area of activated charcoal, soya hull activated carbon, and *L. camara* activated carbon was estimated by Brunauer–Emmett–Teller (BET) method using surface area analyzer (model: Smart Sorb 92/93).

5.2.5 ADSORPTION KINETICS: FREUNDLICH ISOTHERM MODEL

Freundlich isotherm model is a well-known model which describes the non-ideal and reversible adsorption but is not restricted to the formation of monolayer. This empirical model is applicable to multilayer adsorption with uneven distribution of adsorption heat and affinities over the heterogeneous surface.[15] Presently, Freundlich isotherm model is extensively used in heterogeneous systems particularly for organic compounds or extremely interactive species on activated carbon and molecular sieves. The empirical correlation of Freundlich isotherm is given by following equations:

$$q_e = K_f C_e^{1/n} \quad (5.3)$$

$$\ln q_e = \ln K_f + \frac{1}{n} \ln C_e \quad (5.4)$$

K_f is the constant related to adsorption capacity and $1/n$ is the constant related to the adsorption intensity. The value of n suggests the effectiveness of adsorption and if the values of $n > 1$, that represent favorable nature of adsorption.

5.3 RESULTS AND DISCUSSION

5.3.1 FTIR ANALYSIS OF ACTIVATED CHARCOAL, ACTIVATED CARBONS DERIVED FROM LANTANA CAMARA AND SOYA HULL

The FTIR spectrum of activated charcoal and activated carbons derived from *L. camara* and soya hull is depicted in Figure 5.1. The FTIR analysis was carried out in order to establish the probable contribution of the functional groups on the surface of adsorbent during the adsorption of heavy metals. In case of FTIR spectrum of activated charcoal, the broad band at 3132 and 3732 cm^{-1} can be due to —OH stretching vibration (cellulose-based materials) while the band at 2412 cm^{-1} can be allocated to the C≡C and/or C≡N stretching vibrations (amine groups). The band at 3132 and 3732 cm^{-1} are also attributed to a carbonyl group stretching vibrations which are superimposed with hydroxyl group band and that has been reported to fall in the range of 3000–4000 cm^{-1}. The peaks at 1402 and 1560 cm^{-1} may be due to C=O stretching, whereas —CH bending vibrations of carbonyl compound can be attributed to methylene groups. The FTIR spectrum of soya hull activated carbon is also reported in Figure 5.1. The broad band at 3179 cm^{-1} can be assigned to —OH stretching vibration (soya hulls are cellulose-based materials) while the band at 2425 cm^{-1} can

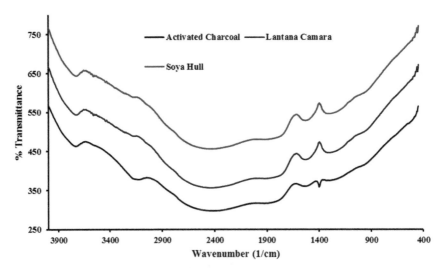

FIGURE 5.1   FTIR spectrum of activated charcoal and activated carbons derived from *Lantana camara* and soya hull.

be attributed to the C≡C and/or C≡N stretching vibrations (amine groups). The band at 3179 cm⁻¹ was also assigned to a carbonyl group stretching vibrations superimposed with hydroxyl group band. The peaks at 1402 and 1634 cm⁻¹ may be because of C=O stretching, and −CH bending vibrations of carbonyl compound can be attributed to methylene groups. Further, the FTIR spectrum of activated carbon derived from *L. camara* is depicted in Figure 5.1. The characteristics peaks at 3158 and 3731 cm⁻¹ are assigned to −OH stretching vibration (*L. camara* stems are cellulose-based materials), whereas, the characteristics peak at 2445 cm⁻¹ is attributed to the C≡C and/ or C≡N stretching vibrations (amine groups). The characteristics peaks at 3158 and 3731 cm⁻¹ were also due to carbonyl group stretching vibrations which are superimposed with hydroxyl group band. Further, the characteristics peak at 1302 and 1535 cm⁻¹ are attributed to C=O stretching, and −CH bending vibrations of carbonyl compound are assigned to methylene groups.

5.3.2 BET CHARACTERIZATION

The specific surface area of activated charcoal, soya hull activated carbon, and *L. camara* activated carbon was estimated by BET method. The BET surface area found for activated charcoal was 955.39 m²/g, for soya hull activated carbon it was 458.54 m²/g, and for *L. camara* activated carbon, it was 689.65 m²/g. It has been observed that with the use of reported method for the preparation of activated carbon from soya hull and *L. camara* stems, the surface area is significantly higher but it is less than commercial activated charcoal.

5.3.3 EFFECT OF INITIAL CONCENTRATION OF HEAVY METAL SOLUTION ON REMOVAL OF Cu²⁺ AND Fe²⁺

The initial solute concentration plays an important role in the adsorption of heavy metal ions on various types of adsorbents. The effect of initial concentration of Cu²⁺ and Fe²⁺ on its percentage removal is depicted in Figures 5.2–5.7. In this study, the initial solute concentration of Cu²⁺ and Fe²⁺ was varied between 10 and 50 ppm. In all the cases, an increase in the percentage removal of Cu²⁺ and Fe²⁺ was observed with an increase in the initial concentration of Cu²⁺ and Fe²⁺. Further, the percentage removal of

Cu^{2+} and Fe^{2+} ions by the prepared adsorbent was found increased rapidly initially with an increase in its concentration and then it slowed down. It is attributed to lesser collisions between metal ion and adsorbent active sites at low concentration of metal ions. However, more collisions between metal ion and adsorbent active sites are possible at higher concentration of metal ions, which leads to increase in the adsorption of heavy metals that results in an increase in percentage removal of Cu^{2+} and Fe^{2+} ions. Also, the rate of adsorption was found to be decreased with an increase in the concentration of Cu^{2+} and Fe^{2+} ions after 20 ppm. This might be due to involvement of less energetically favorable sites with increasing Cu^{2+} and Fe^{2+} ion concentrations in the aqueous solution. Figures 5.2–5.4 show the effect of the initial solute concentration of Cu^{2+} ions on percentage removal of Cu^{2+} using activated charcoal, soya hull activated carbon, and *L. camara* activated carbon at a constant contact time (i.e., 30 min), respectively, at different adsorbent dosage. The percentage removal of Cu^{2+} ions at 10 ppm ion concentration and 0.6 g adsorbent was 50% for all the adsorbent used in this study. It is found to increase with an increase in the initial concentration of Cu^{2+} ions. In the case Fe^{2+} removal, same trend was observed (Figs. 5.5–5.7).

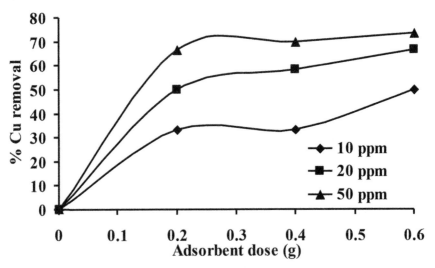

FIGURE 5.2 Effect of activated charcoal loading on Cu removal at different initial concentration (batch time = 30 min).

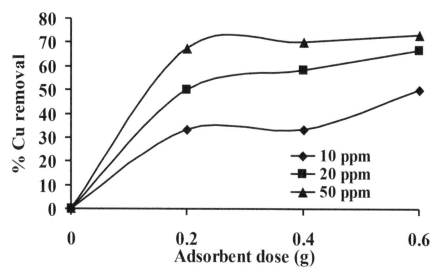

FIGURE 5.3 Effect of soya hull activated carbon loading on Cu removal at different initial concentration (batch time = 30 min).

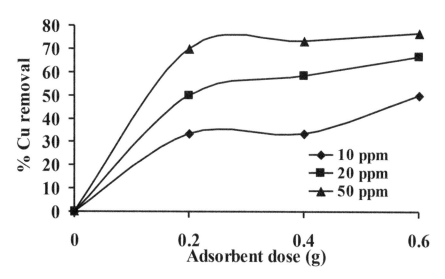

FIGURE 5.4 Effect of *Lantana camara* activated carbon loading on Cu removal at different initial concentration (batch time = 30 min).

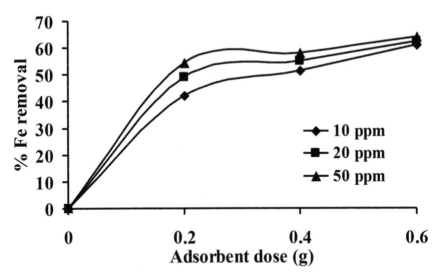

FIGURE 5.5 Effect of activated charcoal loading on Fe removal at different initial concentration (batch time = 30 min).

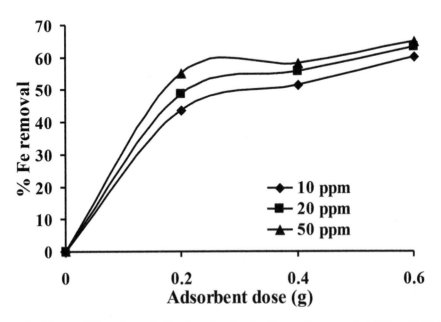

FIGURE 5.6 Effect of soya hull activated carbon loading on Fe removal at different initial concentration (batch time = 30 min).

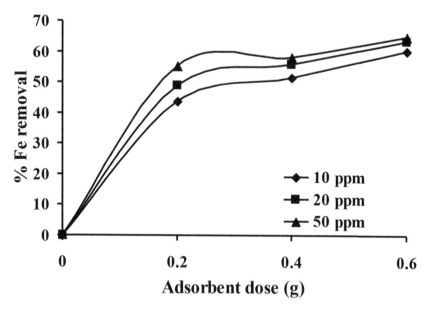

FIGURE 5.7 Effect of *Lantana camara* activated carbon loading on Fe removal at different initial concentration (batch time = 30 min).

5.3.4 EFFECT OF ADSORBENT LOADING ON REMOVAL OF Cu^{2+} AND Fe^{2+}

Adsorbent dose has a great influence on heavy metal removal. Figures 5.2–5.4 and 5.5–5.7 depict the effect of adsorbent dose of various adsorbents at various initial metal ion concentrations. It was found that in all cases, the percentage removal of heavy metal ions increase with an increase in the adsorbent dose. It is attributed to an increase in the removal efficiency of heavy metal with an increase in the number of available active adsorption sites. Amount of adsorbent added to the solution gives the extra number of binding sites which are available for the adsorption. For 30 min contact time, 50 ppm initial concentration and 0.6 g dose of activated charcoal, *L. camara* activated carbon and soya hull activated carbon, the percentage removal of Cu^{2+} was 73.34, 76.68, and 73.00%, respectively. However, for the same condition, the percentage removal of Fe^{2+} was 64.06, 64.00, and 64.84%, respectively. It is observed that the percentage removal of Cu^{2+} and Fe^{2+} using *L. camara* activated carbon and soya hull activated carbon is nearly same to that of activated charcoal. It is attributed to significantly large surface area of *L. camara* activated carbon and soya hull activated carbon, which allows more adsorption of heavy metals on it.

5.3.5 ADSORPTION KINETICS

Freundlich adsorption isotherm studies have been carried out using experimental set of results for all adsorbents and both heavy metal solutions. The constants estimated from the isotherm graphs are reported in Table 5.1. The Freundlich adsorption isotherms for Cu^{2+} and Fe^{2+} using activated charcoal, *L. camara* activated carbon, and soya hull activated carbon are depicted in Figures 5.8 and 5.9. From the R^2 values, it can be predicted that the Freundlich adsorption isotherm is suitable for this study. Further, the value of "n" for various cases ranges from 0 to 1 which is an indication of a favorable condition for adsorption. For all the adsorbents used in this study, the values of constants show better adsorption capacities. This is attributed to higher BET surface area of prepared adsorbent.

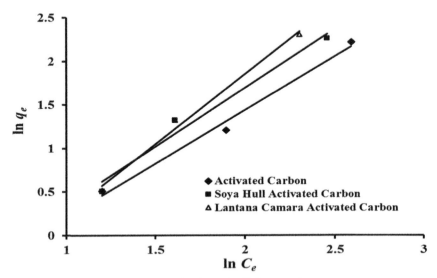

FIGURE 5.8   Freundlich adsorption isotherm for Cu^{2+} removal.

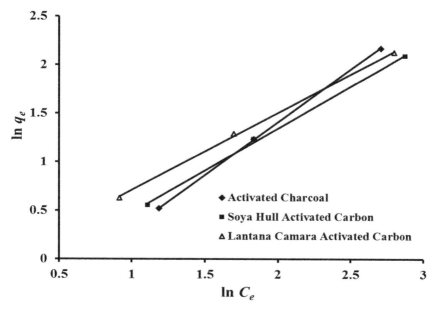

FIGURE 5.9 Freundlich adsorption isotherm for Fe²⁺ removal.

TABLE 5.1 Freundlich Constants for the Adsorption of Cu²⁺ and Fe²⁺.

Type of adsorbent	Cu²⁺ Removal			Fe²⁺ Removal		
	K_f	n	R^2	K_f	n	R^2
Activated charcoal	0.361	0.81	0.9886	0.917	0.46	0.9999
Soya hull activated carbon	0.364	0.74	0.9753	1.151	0.68	0.9985
Lantana camara activated carbon	0.257	0.62	0.9911	1.258	0.92	0.9991

5.4 CONCLUSION

The prepared activated carbon from the various agricultural wastes (i.e., soya hull and *L. camara* stem) has been successfully used in this study for the removal of Fe²⁺ and Cu²⁺ heavy metal ions from the wastewater. The removal efficiency of the prepared adsorbent was well controlled by adsorbent concentration and initial ion concentration. The obtained adsorption results were fitted with Freundlich isotherms. Freundlich isotherm showed better fit with the adsorption data reported in this study. The value of "*n*," that is, Freundlich constant, for various cases ranges from 0 to 1 which is an indication of a favorable condition for adsorption.

KEYWORDS

- agricultural waste
- activated carbon
- soya hull
- *Lantana camara*
- Freundlich isotherm

REFERENCES

1. AL-Othman, Z. A.; Ali, R.; Naushad, M. Hexavalent Chromium Removal from Aqueous Medium by Activated Carbon Prepared from Peanut Shell: Adsorption Kinetics, Equilibrium and Thermodynamic Studies. *Chem. Eng. J.* **2012**, *184*, 238–247.

2. AL-Othman, Z. A.; Naushad, M. Inamuddin, Organic–Inorganic Type Composite Cation Exchanger Poly-O-Toluidine Zr(IV) Tungstate: Preparation, Physicochemical Characterization and its Analytical Application in Separation of Heavy Metals. *Chem. Eng. J.* **2011**, *172*, 369–375.

3. AL-Othman, Z. A.; Naushad, M.; Nilchi, A. Development, Characterization and Ion Exchange Thermodynamics for a New Crystalline Composite Cation Exchange Material: Application for the Removal of Pb^{2+} Ion from a Standard Sample (Rompin Hematite). *J. Inorg. Organomet. Polym. Mater.* **2011**, *21*, 547–559.

4. Bilal, M.; Shah, J. A.; Ashfaq, T.; Gardazi, S. M. H.; Tahir, A. A.; Pervez, A.; Haroon, H.; Mahmood, Q. Waste Biomass Adsorbents for Copper Removal from Industrial Wastewater-A Review. *J. Hazard. Mater.* **2013**, *263*, 322–333.

5. Wang, J.; Chen, C. Biosorbents for Heavy Metals Removal and their Future. *Biotech. Adv.* **2009**, *27*, 195–226.

6. Akar, S. T.; Akar, T.; Kaynak, Z.; Anilan, B.; Cabuk, A.; Tabak, Ã. Z.; Demir, T. A.; Gedikbey, T. Removal of Copper (II) Ions from Synthetic Solution and Real Wastewater by the Combined Action of Dried *Trametes versicolor* Cells and Montmorillonite. *Hydrometallurgy.* **2009**, *97*, 98–104.

7. Andrews, N. C. Disorders of Iron Metabolism. *N. Engl. J. Med.* **1999**, *341*, 1986–1995.

8. Fu, F.; Wang, Q. Removal of Heavy Metal Ions from Wastewaters: A Review. *J. Environ. Manage.* **2011**, *92*, 407–418.

9. Grassi, M.; Kaykioglu, G.; Belgiorno, V.; Lofrano, G. Removal of Emerging Contaminants from Water and Wastewater by Adsorption Process. In *Emerging Compounds Removal from Wastewater, SpringerBriefs in Green Chemistry for Sustainability;* Lofrano, G., Ed.; Springer: New York, 2012; Vol. 2, pp 15–37.

10. Miretzky, P.; Saralegui, A.; Cirelli, A. F. Simultaneous Heavy Metal Removal Mechanism by Dead Macrophytes. *Chemosphere.* **2006**, *62*, 247–254.

11. Yu, X. B.; Wei, C. H.; Ke, L.; Wu, H. Z.; Chai, X. S.; Hu, Y. Preparation of Trimethylchlorosilane-Modified Acid Vermiculites for Removing Diethyl Phthalate From Water. *J. Colloid Interface Sci.* **2012**, *369*, 344–351.

12. Alslaibi, T. M.; Abustan, I.; Ahmad, M. A.; Abu Foul, A. A Review: Production of Activated Carbon from Agricultural Byproducts Via Conventional and Microwave Heating. *J. Chem. Technol. Biotechnol.* **2012,** *88,* 1183–1190.

13. Duman, G.; Onalt, Y.; Okutucu, C.; Onenc, S.; Yanik, J. Production of Activated Carbon from Pine Cone and Evaluation of its Physical, Chemical, and Adsorption Properties. *Energy Fuel.* **2009,** *23,* 2197–2204.

14. Bhatnagar, A.; Vilar, V. J. P.; Botelho, C. M. S.; Boaventura, R. A. R. Coconut-Based Biosorbents for Water Treatment – A Review of the Recent Literature. *Adv. Colloid Interface Sci.* **2010,** *160,* 1–15.

15. Foo, K. Y.; Hameed, B. H. Insights into the Modeling of Adsorption Isotherm Systems. *Chem. Eng. J.* **2010,** *156,* 2–10.

CHAPTER 6

DEFLUORIDATION OF DRINKING WATER USING Fe-Al MIXED METAL HYDROXIDES

T. TELANG[1], M. P. DEOSARKAR[1,*], R. SHETTY[2], and S. P. KAMBLE[2,*]

[1]*Department of Chemical Engineering, Vishwakarma Institute of Technology, Upper Indira Nagar, Bibwewadi, Pune 411037, Maharashtra, India*

[2]*Chemical Engineering and Process Development Division, CSIR-National Chemical Laboratory, Dr. Homi Bhabha Road, Pune 411008, Maharashtra, India*

Corresponding author. E-mail: sp.kamble@ncl.res.in, mpdeosarkar@yahoo.com

CONTENTS

ABSTRACT

In this chapter, we report findings from study carried out on removal of fluoride from drinking or potable water using Fe-Al mixed metal hydroxides. For experimental analysis, the product mixed Fe-Al hydroxides was prepared by applying a co-precipitation method. The resulting metal hydroxide was thoroughly characterized using analytical tests of X-ray diffraction (XRD), scanning electron microscopy (SEM), and Brunauer–Emmett–Teller (BET) surface area analysis. In order to characterize the equilibrium behavior of adsorbent, we study experimentally the effects of different operating parameters. The selected parameters for analysis are adsorbent loading, initial fluoride concentration, pH of the solution, and interfering ions (usually present in groundwater). The equilibrium adsorption data was later fitted for Langmuir adsorption isotherm, and experimental observations show a maximum fluoride adsorption capacity of 2.05 mg.g^{-1}. The experimental results reveal that the adsorption dynamics follows a pseudo-second-order kinetic model.

6.1 INTRODUCTION

It is seen that the fluoride content is a necessary element for normal mineralization of bone and prevention of dental caries. However, any excess intake of fluoride through contaminated drinking water can lead to fluorosis. This is perhaps the most common drinking water related problem. Besides affecting the dental and skeletal system, the fluoride can give rise to various other adverse health effects such as muscle fiber degeneration, low hemoglobin level, deformities in red blood corpuscles (RBCs), excessive thirst, gastrointestinal discomfort, etc. Such effects are also observed due to the excess fluoride intake.[1–3] Over past few decades, the scientific literature on fluoride research suggests that fluoride concentrations between 0.7 and 1.2 mg·L^{-1} are beneficial; and especially so to infants for the prevention of dental caries or tooth decay. However, the fluoride concentrations above 1.2 mg·L^{-1} may cause mottling of teeth in mild cases.

The problem of excess fluoride in drinking water seems to be prevailing in about 25 nations across the world. Further, it is estimated that over 200 million people are bound to consume fluoride-contaminated water across the globe. In India, about 20 out of 35 states and Union Territories are affected due to the excess fluoride in groundwater covering more than 150 districts and 36,988 habitations.[4] Clearly, as we consider the severity of health problems due to the excess fluoride in groundwater, the World Health Organization (WHO) has reduced the permissible limit of fluoride from 1.5 to 1.02 mg·L^{-1} in 1998. Furthermore, due to harm inflicted by excess fluoride content to human health, we see that when fluoride content overshoots a critical level, it becomes necessary to reduce and control the fluoride concentration in public water supply systems.[5–8]

Literature study shows that there are various methods that were used to remove fluoride from water, which include adsorption,[9,10] precipitation,[11,12] ion exchange,[13] membrane processes,[14] electro dialysis,[15] reverse osmosis,[16] and nanofiltration. Among these methods, adsorption is the most suitable and widely used technique due to its simple operation, and the availability of a wide range of adsorbents.[17–19]

The criteria for the selection of adsorbent include its capacity for fluoride removal and product cost.[20] Until now, various low-cost materials such as activated alumina,[21] clay,[22] soil,[23] bone char,[24] light weight concrete,[25] fly ash, and other materials[26] have been tested for removing fluoride from drinking water. But the fluoride adsorption capacities of these adsorbents were found not high enough for potential use for commercialization. In the past few years, a number of novel adsorbents with strong affinity toward fluoride have been developed for fluoride removal. These adsorbents include calcined Mg–Al–CO$_3$ layered double hydroxides (LDH),[27] zirconium-impregnated collagen fiber,[28] Fe–Al–Ce trimetal oxide,[29] and iron-aluminum mixed oxide.[30] Also, rare earth elements have been tried as potential adsorbents because of their selective affinity to fluoride, high adsorption capacity; we further note that these adsorbents cause little pollution and easy preparation.[31] These novel adsorbents were found to possess substantial potential for fluoride adsorption; but some of these adsorbents were expensive and may not have enough economic viability to design industrial equipment or products for large-scale drinking water treatment facility. Therefore, we should be looking for an effective and low-cost adsorbent that will have high-fluoride adsorption capacity, and that can become efficient water treatment technology available for removal of fluoride is desirable.

In this chapter, we will focus upon the mixed Fe/Al based materials for fluoride removal. The Fe/Al mixed hydroxides were prepared by simple precipitation method, and characterized using X-ray diffraction (XRD), scanning electron microscopy (SEM), energy dispersive analysis of X-rays (EDAX), and Brunauer–Emmett–Teller (BET) surface area instruments. The effect of various parameters such as adsorbent dose, initial fluoride concentration, contact time, pH of solution, and co-existing ions on uptake of fluoride ions have been studied in detail. The nature of this adsorption system was described by kinetics as well as column study. The most common adsorption isotherm namely, Langmuir, and Freundlich isotherm models were tested to observe the equilibrium behavior of the system.

6.2   MATERIALS AND METHODS

6.2.1   MATERIALS

To carry out the experimental synthesis, anhydrous iron (III) chloride (M/s Merck India Ltd., Mumbai, India), aluminum nitrate nanohydrate (Thomas Baker, Mumbai, India), sodium hydroxide, and sodium carbonate were used for synthesis of adsorbents. Also, sodium chloride, sodium sulfate, sodium nitrate, sodium carbonate, and sodium hydroxide were used to study the effect of various parameters discussed above on adsorption isotherm and were purchased from M/s Merck India Ltd., Mumbai, India. Fluoride solution used in experiments was prepared by dissolving sodium fluoride (NaF) in de-ionized (DI) water. Note that all chemical reagents used in this study were of analytical grade.

6.2.2   SYNTHESIS

For the synthesis of Fe-Al mixed metal hydroxides two solutions were prepared. Further, solution A was prepared by mixing 64.88 g, 2 M $FeCl_3$ and 74.026 g, 1 M $Al(NO_3)_3 \cdot 9H_2O$ in 200 mL DI water. Solution B was prepared by dissolving sodium hydroxide (0.5 N) and sodium carbonate (0.2 N) in 200 mL of DI water. These two solutions were added drop wise to a beaker containing 100 mL of DI water with continuous stirring for 4 h for complete addition and mixing until pH of the slurry was 10. The slurry was filtered using vacuum filter and the precipitate was washed twice using DI water. The precipitate was then dried in oven at 110°C overnight. The sample was calcined for 500°C for 3 h.

6.2.3   ADSORPTION EXPERIMENT

Fluoride solution (100 ppm) was prepared by dissolving 2.21 mg of NaF in 1000 mL of DI water as a stock solution. A known weight of adsorbent material (Fe-Al mixed metal hydroxides) was added to fluoride solution in a conical flask and kept on a rotary shaker (Serial No.RIS-246, Remi Instruments, Mumbai, India) for 24 h at a constant speed of 150 rpm. After 24 h, the solution was filtered through Whitman filter paper 125 mm Ø under vacuum. A 10 mL portion of the filtrate was taken in a beaker and 1 mL total ionic strength adjustment buffer (TISAB) III solution was added for pH adjustment. Fluoride ion analysis was then performed by using fluoride Orion-ion selective electrode. Batch adsorption experiments were conducted to investigate the effect of various parameters like adsorbent dose, initial concentration, presence of interfering ions, pH, etc. The specific amount of fluoride adsorbed was calculated using the following equation

$$q_e = (C_o - C_e) \times \frac{V}{W} \tag{6.1}$$

where q_e is the adsorption capacity (mg·g^{-1}) in the solid at equilibrium; C_o, C_e are initial and equilibrium concentration of fluoride (mg·L^{-1}), respectively; V is the volume of the aqueous solution (L) and W is the mass (g) of adsorbent used in the experiments.

The fluoride removal efficiency was calculated using the expression:

$$RE(\%) = \frac{(C_o - C_e)}{C_o} \times 100 \tag{6.2}$$

where, RE (%) is the fluoride removal efficiency; C_o and C_e are the initial and the equilibrium fluoride concentration (mg·L^{-1}), respectively.

6.2.4 KINETIC EXPERIMENTS

In order to estimate equilibrium adsorption rate for the uptake of fluoride by Fe-Al mixed metal hydroxides, time-dependent sorption studies were conducted in a polyvinyl chloride (PVC) vessel having a capacity of 1000 mL. The particular concentration of fluoride solution was transferred into the vessel and a known weight of the adsorbent was added. The suspension was stirred using a four-blade, pitched turbine impeller with the stirring speed up to 200 rpm. Samples were withdrawn from the vessel at frequent time intervals of 1 min and filtered through Whitman filter paper of 125 mm Ø. The filtrate was analyzed for the fluoride concentration using an Orion ion-selective electrode.

6.2.5 COLUMN EXPERIMENTS

The adsorption column was made of stainless steel (SS) having diameter of 1 cm and length 25 cm. The top and bottom connections were made using ferrule fitting to prevent leakage of liquid during experiments. The SSI pump (supplied by M/s Texol Pvt. Ltd, Pune, India) was used to pass the fluoride solution. The column was packed by Fe-Al mixed metal hydroxides + sand mixture (50% Fe-Al mixed metal hydroxides and 50% sand). The packing material, glass bead, and glass wool were stuffed at the top and bottom of about 2.5 cm. Sand is used to reduce the pressure drop and increase voltage since the adsorbent is having a very fine mesh size. The outlet solution was filtered using Whitman filter paper and the filtrate was analyzed.

6.3 RESULTS AND DISCUSSIONS

6.3.1 CHARACTERIZATION OF THE FE-AL MIXED METAL HYDROXIDES

Chemical composition of the Fe-Al mixed metal hydroxides was determined by XRD. Figure 6.1 shows that the XRD pattern for Fe-Al hydroxide before and after adsorption of fluoride in aqueous solution. The dominant peaks $2\theta = 27.8, 33.1$ of before adsorption confirm the presence of hematite and aluminum in bayerite structure. XRD pattern also reveals that the crystalline structure of material shows significant changes after adsorption of fluoride ions. Further, it is observed that there were significant changes in peak position after fluoride adsorption with $2\theta = 16.1, 22.3, 26,$ and 37.4 indicating the formation of AlF_3 and FeF_3. The results obtained here in an agreement with similar studies reported earlier.[32,33] The adsorbent completely changes its structure after adsorption. This suggests that the uptake of fluoride ions is partly by chemical adsorption. The surface area of the adsorbent was analyzed by single point BET method and was found to be 72.40 $m^2 \cdot g^{-1}$ and their total pore volume was 0.2483 $cc \cdot g^{-1}$.

Figure 6.2 shows SEM images before adsorption and clearly reveals the floc, porosity, and surface texture, which has endowed the adsorbents with large surface areas and high adsorption capacity. The size of particle was found to be about 5 μm. Figure 6.2b shows less porous structure of the material after adsorption. The reduction in the porosity of the material confirms the adsorption of fluoride on the material.

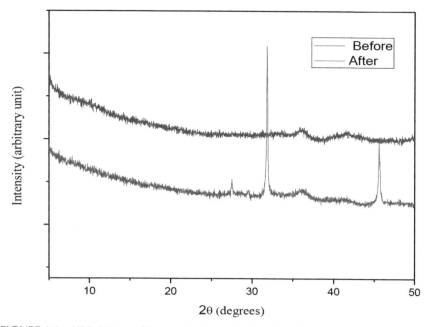

FIGURE 6.1   XRD Pattern of Fe-Al mixed metal hydroxide before and after adsorption.

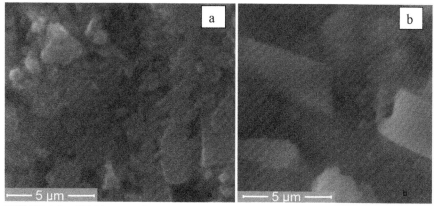

FIGURE 6.2   SEM images of Fe-Al mixed metal hydroxide. (a) Before adsorption. (b) After adsorption.

The elemental composition by atomic and weight percent of the adsorbent was determined by quantitative analysis of energy-dispersive X-ray spectroscopy (EDX). Figure 6.3a shows the weight percent of 25.78, 8.81, 1.13, and 65.46 for O, Al, Cl, and Fe. Figure 6.3b shows the fluoride-loaded sample with weight percent of 32.37, 2.31, 0.44, 0.97, 15.47, and 50.39 for O, F, Na, Al, and Fe, respectively.

The EDX analysis of the material (see Fig. 6.3a), before adsorption does not shows any presence of fluoride; but after adsorption (see Fig. 6.3b), presence of fluoride is observed. And then it clearly indicates that the fluoride has been absorbed by the adsorbent.

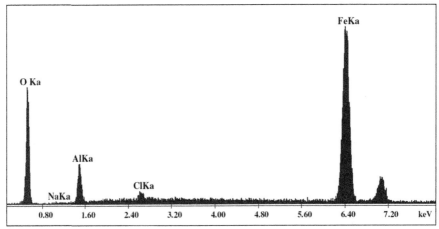

FIGURE 6.3a   EDX of Fe-Al mixed metal hydroxide before adsorption.

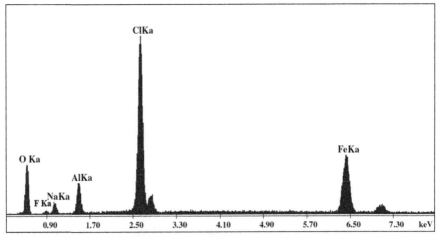

FIGURE 6.3b   EDX of Fe-Al mixed metal hydroxide after adsorption.

6.3.2 EFFECT OF ADSORBENT DOSE

Figure 6.4 shows the effect of adsorbent dose upon uptake of the fluoride at fixed initial fluoride concentration, pH 7, shaker speed of 150 rpm, and contact time of 24 h. The adsorbent dose was varied from 0.1 to 1.2 g. The results show that the adsorption capacity decreases from 1.27 to 0.18 mg·g⁻¹. An increase in the mass of adsorbent at a fixed fluoride concentration leads to a decrease in q_e, because there are too many adsorbent particles for the limited amount of fluoride. We can therefore conclude that the fluoride removal efficiency increases with the adsorbent dose due to more active sites.

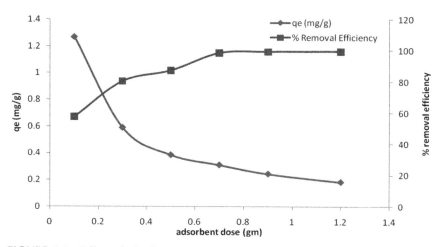

FIGURE 6.4   Effect of adsorbent dose on Fe-Al mixed metal hydroxide (initial fluoride concentration = 4.41 mg·L⁻¹, contact time = 24 h, pH of solution = 7, and shaker speed = 150 rpm).

6.3.3 EFFECT OF INITIAL FLUORIDE CONCENTRATION

Figure 6.5 shows the effect of initial fluoride concentration on adsorption capacity. The effect of initial concentration on percentage removal of fluoride was studied by varying initial fluoride concentration, while keeping the other parameters constant. Figure 6.5 shows that the fluoride removal capacity of adsorbent increases with the increase in initial concentration of fluoride content, which may be due to more availability of adsorbate, and that the active sites of the adsorbent have not reached saturation even at initial fluoride concentrations of 20 mg·L^{-1}.

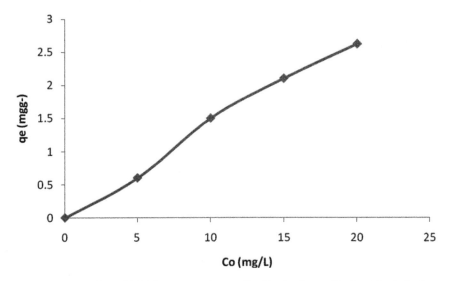

FIGURE 6.5   Effect of initial concentration on Fe-Al mixed metal hydroxides (adsorbent dose = 0.3 g/50 mL, contact time = 24 h, pH of solution = 7, and shaker speed = 150 rpm).

6.3.4   EFFECT OF pH

The pH of the solution plays an important role and controls the adsorption at water-adsorbent interface. Also, the pH of the drinking water predominantly depends upon geological characteristics of soil as well as weather conditions. Therefore, it becomes necessary to study the effect of pH on uptake of fluoride. In accordance with the guidelines of WHO the normal range of pH in drinking water is 6–8.5. The pH of fluoride solution was adjusted using 0.1 N HCl and 0.1 N NaOH solutions. T he adsorption of fluoride was carried out at different pH ranging from 3, 5, 7, 9 to 11 by keeping other parameters constant. The uptake of fluoride by Fe-Al mixed metal hydroxides over the pH range of 3–11 is shown in Figure 6.6. It is apparent from experimental observations that the pH has some effect on fluoride removal efficiency. And that the fluoride removal efficiency increases with the increase of pH, and it reaches a maximum value of 80.85% at pH 11.

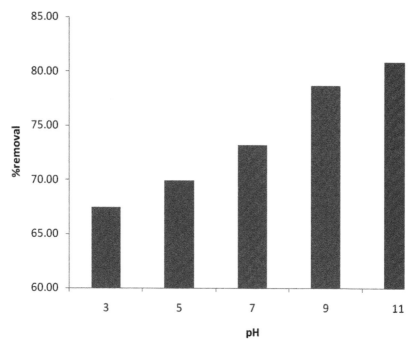

FIGURE 6.6 Effect of pH on Fe-Al mixed metal hydroxides (initial concentration of fluoride solution = 9.57 mg·L⁻¹, dose of adsorbent = 0.3 g/50mL, contact time = 24 h, adsorption temperature = 30±2 °C, and constant shaker speed = 150 rpm).

6.3.5 EFFECT OF THE PRESENCE OF CO-EXISTING IONS

Drinking water and wastewater contain several other anions that might affect the adsorption process. In order to understand the effect of interfering ions, adsorption studies were carried out in the presence of sodium chloride, sodium sulfate, sodium nitrate, and sodium carbonate for adsorption of fluoride. It may be noted that initial fluoride concentration was 9.57 mg·L⁻¹, shaker speed of 150 rpm, and contact time of 24 h. The salt concentrations were taken as 100 mg·L⁻¹ in order to interpret the adsorption trend of fluoride even in the presence of these ions. Figure 6.7 shows the effect of different concentrations of anions on the uptake of fluoride ions from the aqueous solution. Figure 6.7 shows that there is a significant decrease in the removal of fluoride due to the co-existing ionic interference. The maximum decrease in fluoride removal of around 20% has been observed for NaCl ions.

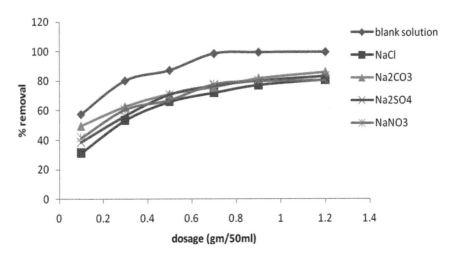

FIGURE 6.7   Effect of presence of co-existing ions (initial concentration of fluoride solution = 9.57 mg·L^{-1}, contact time = 24 h, adsorption temperature = 30 ± 2°C, and constant shaker speed = 150 rpm).

6.3.6 EQUILIBRIUM MODELING

In this chapter, the Langmuir and Freundlich isotherm models were used to obtain various adsorption parameters and in order to get an insight into the adsorption mechanism.

The Freundlich model is an empirical equation based on sorption on a heterogeneous surface. The expression is given as:

$$q_e = K_f C_e^{1/n} \qquad (6.3)$$

where K_f and n are the Freundlich constants that indicate relative capacity and adsorption intensity, respectively. The linear form of the Freundlich equation (refer to eq 6.4) is commonly used to fit experimental data to this adsorption isotherm.[34]

$$\log q_e = \log K_f + \frac{1}{n}\log C_e \qquad (6.4)$$

The experimental data of the present works were tested with the Freundlich model (given in eq 6.4). The plot of log (q_e) vs log (C_e) prepared using eq 6.4 is found to be a straight line as clearly seen from Figure 6.8. The values of K_f and $1/n$ for Fe-Al mixed metal hydroxide were 0.696 and 0.214 mg·g^{-1}, respectively, for initial concentration of fluoride solution of 4.57 mg·L⁻¹. Since the value of the constant, $1/n$ (that gives adsorption intensity) is less than unity, and this clearly confirms a satisfactory extent of adsorption.

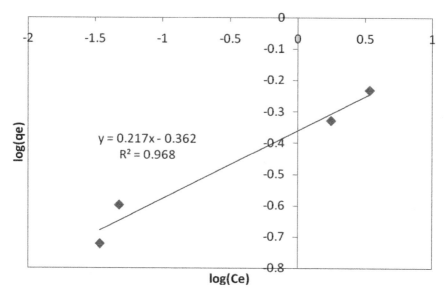

FIGURE 6.8 Freundlich isotherm for Fe-Al mixed metal hydroxides (volume = 50 mL, initial fluoride concentration = 4.57 mg·L⁻¹, and contact time = 24 h).

The Langmuir equation which is valid for monolayer sorption onto a surface is

$$\frac{1}{q_e} = \frac{1}{q_{max}K} \times \frac{1}{C_e} + \frac{1}{q_{max}} \tag{6.5}$$

where q_e is the amount of fluoride ions adsorbed per unit weight of Fe-Al mixed metal hydroxide (mg·g⁻¹), q_{max} is the maximum sorption capacity corresponding to complete monolayer coverage (mg·g⁻¹), Langmuir constant K is indirectly related to the energy of adsorption (L·mg⁻¹), and C_e is the equilibrium concentration of fluoride ion (mg·L⁻¹).

The values of Langmuir parameters, q_{max} and K were calculated from the slope and intercept of the linear plots of $1/q_e$ vs $1/C_e$ (Fig. 6.9) and were found to be 0.524 mg·g⁻¹ and 0.0214 L·mg⁻¹, respectively, that has a regression coefficient (R^2) of 0.97.

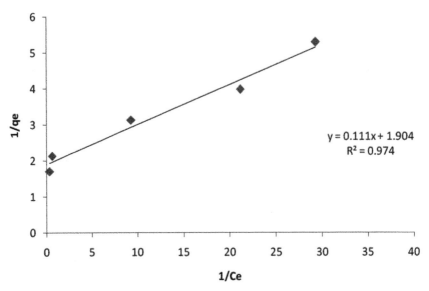

FIGURE 6.9 Langmuir isotherm for Fe-Al mixed metal hydroxides (volume = 50 mL, initial fluoride concentration = 4.57 mg·L^{-1}, and contact time = 24 h).

In order to predict the adsorption efficiency of the process, the dimensionless quantity (r) was calculated by using the following equation[35,36]

$$r = \frac{1}{1 + KC_o} \qquad (6.6)$$

where K is a constant in Langmuir isotherm and C_o is initial concentration of fluoride. If the value of $r < 1$, it represents favorable adsorption, while >1.0 represents unfavorable adsorption. The value of r for an initial fluoride concentration of 4.57 mg·L^{-1} was found to be 0.0129. It indicates that the system is favorable for adsorption.

6.3.7 KINETIC ADSORPTION ON FE-AL MIXED METAL HYDROXIDE ADSORBENT

Adsorption kinetics is one of the most important characters which represent the adsorption efficiency. The adsorption rate of fluoride adsorption on the Fe-Al mixed metal hydroxide surface is a function of the initial fluoride concentration, and is shown in Figure 6.10. It was observed that the uptake of fluoride increases with time. However, the adsorption of fluoride was rapid in the first 15 min after which the rate slowed down. The time required to reach equilibrium was 60 min. The pseudo-first-order, pseudo-second-order, and diffusion models were tested to investigate the mechanism of adsorption. The pseudo-first-order rate expression of Lagergren is given as[37,38]

$$\log(q_e - q) = \log q_e - \frac{K_{lad}}{2.303}t \tag{6.7}$$

where q_e and q (both in mg·g^{-1}) are the amounts of fluoride adsorbed per unit mass of adsorbent at equilibrium and at time t, respectively. Also, K_{lad} is the pseudo-first-order rate constant (min^{-1}).

The value of K_{lad} was calculated from the slope of the linear plot of log (q_e-q) vs time (see Fig. 6.10). The adsorption constant were found to be 0.0023, 0.00115, and 0.001612 min^{-1} for initial fluoride concentration of 5.55, 9.88, and 17.5 mg·L^{-1}, respectively.

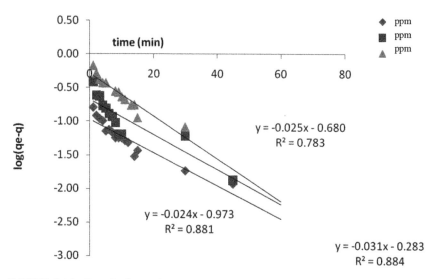

FIGURE 6.10   Pseudo-first-order kinetic plots for the adsorption of fluoride onto Fe-Al mixed metal hydroxide adsorbent at various initial concentrations (adsorbent dose = 6 g·L^{-1}, pH = 7).

The following equation is pseudo-second-order kinetics model tested with the experimental results[37]

$$\frac{t}{q} = \frac{1}{K_{2ads}\, q_e^2} + \frac{1}{q_e} \tag{6.8}$$

where K_{2ads} (g·mg·min^{-1}) is the rate constant for second-order adsorption. The linear equation was used to plot a graph t/q vs t (as shown in Fig. 6.11). The rate constants for the second order reaction with their regression coefficients are shown in Table 6.1. For the pseudo-first-order adsorption the R^2 value is 0.93, whereas for the second-order R^2 value is 0.99 and also equilibrium adsorption capacity calculated from second-order adsorption plot matches with the values obtained in the experiment. Therefore, it can be concluded that the adsorption of fluoride on Fe-Al mixed metal hydroxide follows pseudo-second-order adsorption kinetics.

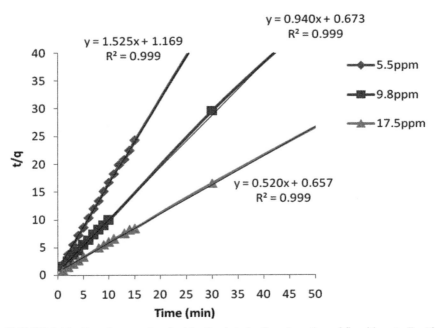

FIGURE 6.11   Pseudo-second-order kinetic plots for the adsorption of fluoride onto Fe-Al mixed metal hydroxide adsorbent at various initial concentrations (adsorbent dose = 6 g·L^{-1}, pH = 7).

TABLE 6.1   Kinetic Parameters for Fluoride Adsorption on Fe-Al Mixed Metal Hydroxides.

C_0 (mg·L^{-1})	q_e exp (mg·g^{-1})	Pseudo-first-order kinetics			Pseudo-second-order kinetics		
		K_{1ad} (min^{-1})	$q_{e(cal)}$ (mg·g^{-1})	R^2	K_{2ads} (g·mg·min^{-1})	$q_{e(cal)}$ (mg·g^{-1})	R^2
5.5	0.61	0.056	0.37	0.88	2.0	0.65	0.99
9.8	1.04	0.073	0.75	0.88	1.06	1.06	0.99
17.5	1.90	0.059	0.5	0.78	0.41	1.92	0.99

We also noted that the adsorption on porous solid adsorbent can also be modeled by pore diffusion models, which can be either particle diffusion or pore diffusion model. The particle diffusion model can be written as[38]

$$In\left(\frac{C_t}{C_e}\right) = -K_p t \qquad (6.9)$$

where K_p is the particle diffusion coefficient (mg·g^{-1}·min). The value of K_p can be obtained by slope of the plot between $In(C_t)$ and t (see Fig. 6.12).

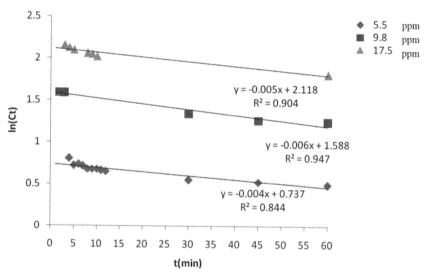

FIGURE 6.12 Particle diffusion plots for the adsorption of fluoride onto Fe-Al mixed metal hydroxide adsorbent at various initial concentrations (adsorbent dose = 6 g·L^{-1}, pH = 7).

The intra-particle pore diffusion model is also commonly used to characterize the sorption data.[39] According to this model, if the rate-limiting step is diffusion of adsorbate within the pores of adsorbent particle (intra-particle diffusion) a graph between amount of adsorbate adsorbed and square root of time should give a straight line passing through the origin. The equation can be written as

$$q_t = K_i t^{1/2} \qquad (6.10)$$

where K_i is the intra-particle diffusion coefficient (mg^{-1}·g·min$^{0.5}$), which can be obtained from the slope of plot of q_t vs $t^{1/2}$. The plots of linear forms of intra-particle pore diffusion model are given in Figure 6.13, and the values of different parameters are given in Table 6.2. The values of R^2 for particle diffusion model are closer to unity indicating that particle diffusion of adsorbate is contributing more toward rate determining step. However, in case of intra-particle diffusion model the lines are not passing through the origin, which reveals that the adsorption of fluoride on Fe-Al mixed metal hydroxides is a complex process involving surface adsorption, inter-particle diffusion, and intra-particle diffusion, all contributing toward the rate of adsorption.

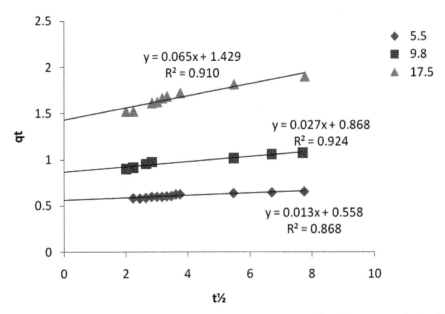

FIGURE 6.13 Intra particle diffusion plots for the adsorption of fluoride onto Fe-Al mixed metal hydroxide adsorbent at various initial concentrations (adsorbent dose = 6 g·L^{-1}, pH = 7).

TABLE 6.2 Diffusion Parameters for Fluoride Adsorption on Fe-Al Mixed Metal Hydroxides.

C_o (mg·L^{-1})	Particle diffusion model		Intra particle diffusion model	
	K_p (min^{-1})	R^2	K_i (min^{-1})	R^2
5.5	0.0046	0.84	0.013	0.86
9.8	0.0066	0.95	0.027	0.92
17.5	0.0053	0.90	0.065	0.91

6.3.8 COLUMN STUDIES

The column study was performed to investigate the feasibility of the adsorbent at domestic/community level. The breakthrough column was used for removal of fluoride ions from drinking water. The effect of various process parameters such as inlet flow rate, initial fluoride concentration, and bed height at various throughput volumes on breakthrough capacity was studied.

6.3.8.1 EFFECT OF FLOW RATE ON FLUORIDE SOLUTION

In this study, the initial concentration of fluoride solution (5 mg·L^{-1}) and bed height (10 cm) was kept constant and only flow rate of the solution through the column was varied. The breakthrough curves are show in Figure 6.14. It was observed that as the flow rate was increased the breakthrough curve becomes sharper, and subsequently the breakthrough time and adsorbed fluoride concentration decreases with an increase in initial flow rate of feed fluoride solution. This is due to the shorter residence time of fluoride ion in column at a higher flow rate. It is expected that at a lower flow rate due to adequate interaction time adsorption was very efficient at least in the initial step of the process, and hence highest breakthrough time. The column can adsorb fluoride after the breakthrough curve, although at a lower efficiency.

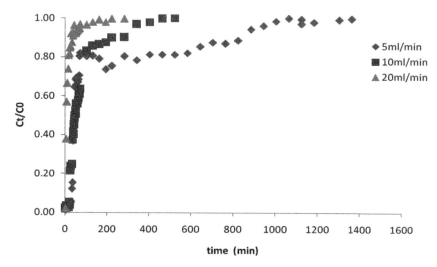

FIGURE 6.14 Effect of flow rate on adsorption of fluoride on breakthrough curve (column height 10 cm and initial concentration 4.11 mg·L^{-1}).

6.3.8.2 EFFECT OF BED HEIGHT

The effect of bed height on breakthrough capacity is shown in Figure 6.15. In this study, initial concentration of fluoride solution (5 mg·L⁻¹) and flow rate (10 cm³·min⁻¹) was kept constant and bed height was varied from 10, 15 and 20 cm. It was observed that as the bed height increases, the breakthrough curve becomes steeper and subsequently capacity also increases. With an increase in bed height, the residence time of solution and the percentage of saturation increases. The breakthrough capacity was found to be 0.3211, 0.3957, and 0.4231 cm³·g⁻¹ for bed height 10, 15, and 20 cm, respectively.

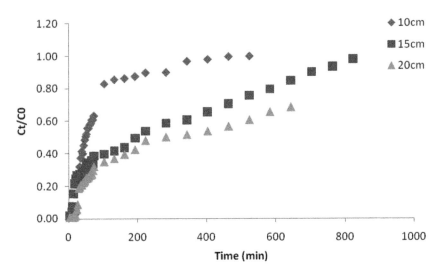

FIGURE 6.15   Effect of bed height on breakthrough curve (initial concentration = 4.11 mg·L⁻¹ and flow rate 10 cm³·min⁻¹).

6.4   CONCLUSION

The Fe-Al mixed metal hydroxide was synthesized for defluoridation by co-precipitation method. The adsorbent showed excellent fluoride removal efficiency of 99.57% and adsorption capacity of 1.27 mg·g⁻¹ at pH 7 for initial fluoride concentration of 5 mg·L⁻¹. The removal of fluoride from aqueous solutions depends on initial fluoride concentration, adsorbent dose, and pH of the solution.

The removal of fluoride ions from water by adsorption process followed the Langmuir and Freundlich adsorption isotherm. But it was best fitted to Freundlich isotherm. As the regression constant for the Freundlich isotherm was higher than the Langmuir isotherm. Kinetic experiments were performed and it was found that the kinetic data fitted well to pseudo-second-order model. The column breakthrough studies were performed and it was found that the breakthrough adsorption capacity is strongly dependent upon the feed flow rate and bed height.

ACKNOWLEDGEMENT

Financial support from CSIR, Government of India under project "Development of Sustainable Waste Management Technologies for Chemical and Allied industries (SETCA) Project # CSC-0113" for this work is gratefully acknowledged.

NOMENCLATURE

C_o	Initial concentration of fluoride $(mg \cdot L^{-1})$
C_e	Equilibrium concentration of fluoride $(mg \cdot L^{-1})$
K_f	Freundlich constant related to adsorption capacity
K	Langmuir constant indirectly related to the energy of adsorption $(L \cdot mg^{-1})$
K_{1ad}	Pseudo-first-order rate constant (min^{-1})
K_{2ads}	Pseudo-second-order rate constant $(g \cdot mg \cdot min^{-1})$
K_p	Particle diffusion coefficient $(mg \cdot g^{-1} \cdot min)$
K_i	Intra-particle diffusion coefficient $(mg^{-1} \cdot g \cdot min^{0.5})$
n	Adsorption intensity
q_e	Adsorption capacity at equilibrium $(mg \cdot g^{-1})$
q_{max}	Maximum sorption capacity corresponding to complete monolayer coverage $(mg \cdot g^{-1})$
RE (%)	Fluoride removal efficiency
t	Time (min)
V	Volume of the aqueous solution (L)
W	Weight of adsorbent used in the experiments (g)

KEYWORDS

- **fluoride**
- **Fe-Al mixed metal hydroxides**
- **adsorption**
- **kinetic modeling**
- **column breakthrough studies**

REFERENCES

1. Sujana, M.; Soma, G.; Vasumathi, N.; Anand, S. Studies on Fluoride Adsorption Capacities of Amorphous Fe/Al Mixed Hydroxides from Aqueous Aolutions. *J. Fluorine Chem.* **2009,** *130,* 749–754.
2. Tripathy, S.; Bersillon, J.; Gopal, K. Removal of Fluoride from Drinking Water by Adsorption onto Alum-Impregnated Activated Alumina. *Sep. Purif. Technol.* **2006,** *50,* 310–317.
3. Bansiwal, A.; Pillewan, P.; Biniwale, R.; Rayalu, S. Copper Oxide Incorporated Mesoporous Alumina for Defluoridation of Drinking. *Micropor. Mesopor. Mat.* **2010,** *129,* 54–61.
4. Ayoob, S.; Gupta, A. K. Fluoride in Drinking Water: A Review on the Status and Stress Effects. *Crit. Rev. Environ. Sci. Technol.* **2006,** *36,* 433–487.
5. World Health Organization (WHO). *Guidelines for Drinking Water Quality;* 3rd ed.; WHO: Geneva, Switzerland, 2004; Vol. 1.
6. Tewari, A.; Dubey, A. Defluoridation of Drinking Water: Efficiency and Need. *J. Chem. Pharma. Res.* **2009,** *1,* 31–37.
7. Dongre, R.; Ghuga, D.; Meshram, J.; Ramteke, D. Fluoride Removal from Water by Zirconium(IV) Doped Chitosan Bio-Composite. *Afr. J. Environ. Sci. Technol.* **2012,** *6,* 130–141.
8. World Health Organization. *Guidelines for Drinking Water Quality Incorporating First Addendum Recommendations;* 3rd ed.; WHO: Geneva, Switzerland, 2006; Vol. 1, pp 375–376.
9. Bulut, Y.; Tez, Z. Adsorption Studies on Ground Shells of Hazelnut and Almond. *J. Hazard. Mat.* **2007,** *149,* 35–41.
10. Veli, S.; Alyuz, B. Adsorption of Copper and Zinc from Aqueous Solutions by Using Natural Clay. *J. Hazard. Mat.* **2007,** *149,* 226–233.
11. Shrivastava, B.; Vani, A. Comparative Study of Defluoridation Technologies in India. *Asian J. Exp. Sci.* **2009,** *23,* 269–274.
12. Chang, M.; Liu, J. Precipitation Removal of Fluoride from Semiconductor Wastewater. *J. Environ. Eng.* **2007,** *133,* 419–425.
13. Castel, C.; Schweizer, M.; Simonnot, M.; Sardin, M. Selective Removal of Fluoride Ions by a Two-Way Ion-Exchange Cyclic Process. *Chem. Eng. Sci.* **2000,** *55,* 3341–3352.

14. Yadav, B.; Garg, A. Impact and Remedial Strategy of Fluoride in Ground Water–A Review. *Int. J. Eng. Res. App.* **2014**, *4*, 570–577.

15. Durmaz, F.; Kara, H.; Cengeloglu, Y.; Ersoz, M. Fluoride Removal by Donnan Dialysis with Anion Exchange Membranes. *Desalination.* **2005**, *177*, 51–57.

16. Ning, R. Arsenic Removal by Reverse Osmosis. *Desalination.* **2002**, *143*, 237–241.

17. Gill, T.; Tiwari, S.; Kumar, P. A Review on Feasibility of Conventional Fluoride Removal Techniques in Urban Areas. *Int. J. Environ. Res. Develop.* **2014**, *4*, 179–182.

18. Mohapatraa, M.; Ananda, S.; Mishraa, B.; Giles, D.; Singh, P. Review of Fluoride Removal from Drinking Water. *J. Environ. Manage.* **2009**, *91*, 67–77.

19. Ayoob, S.; Gupta, A.; Bhat, V. A Conceptual Overview on Sustainable Technologies for Defluoridation of Drinking Water and Removal Mechanisms. *Crit. Rev. Environ. Sci. Technol.* **2008**, *38*, 401–470.

20. Zhaoa, X.; Wanga, J.; Wub, F.; Wanga, T.; Caia, Y.; Shia, Y.; Jianga, G. Removal of Fluoride from Aqueous Media by $Fe_3O_4@Al(OH)_3$ Magnetic Nanoparticles. *J. Hazard. Mat.* **2010**, *173*, 102–109.

21. Ayoob, S.; Gupta, A.; Bhakat, P.; Bhat, V. Investigations on the Kinetics and Mechanisms of Sorptive Removal of Fluoride from Water Using Alumina Cement Granules. *J. Chem. Eng.* **2008**, *140*, 6–14.

22. Hamdi, N.; Srasra, E. Removal of Fluoride from Acidic Wastewater by Clay Mineral: Effect of Solid–Liquid Ratios. *Desalination.* **2007**, *206*, 238–244.

23. Wang, Y.; Reardon, E. Activation and Regeneration of a Soil Sorbent for Defluoridation of Drinking Water. *Appl. Geochem.* **2001**, *16*, 531–539.

24. Castillo, N.; Ramos, R.; Perez, R.; Cruz, R.; Rosales, A.; Coronado, R.; Rubio, L. Adsorption of Fluoride from Water Solution on Bone Char. *Ind. Eng. Chem. Res.* **2007**, *46*, 9205–9212.

25. Oguz, E. Equilibrium Isotherms and Kinetic Studies for the Sorption of fluoride on Light Weight Concrete Materials. *Colloids Surf. A Physicochem. Eng. Asp.* **2007**, *295*, 258–263.

26. Shen, F.; Chen, X.; Gao, P.; Chen, G. Electrochemical Removal of Fuoride Ions from Industrial Waste Water. *Chem. Eng. Sci.* **2003**, *58*, 987–993.

27. Das, J.; Patra, B.; Baliarsingh, N.; Parida, K. Adsorption of Phosphate by Layered Double Hydroxides in Aqueous Solutions. *Appl. Clay Sci.* **2006**, *32*, 252–260.

28. Liao, X.; Shi, B. Adsorption of Fluoride on Zirconium (IV)-Impregnated Collagen Fiber. *Environ. Sci. Technol.* **2005**, *39*, 4628–4632.

29. Wu, X.; Zhang, Y.; Dou, X.; Yang, M. Fluoride Removal Performance of a Novel Fe–Al–Ce Trimetal Oxide Adsorbent. *Chemosphere.* **2007**, *69*, 1758–1764.

30. Biswas, K.; Saha, S.; Ghosh, U. Adsorption of Fluoride from Aqueous Solution by a Synthetic Iron(III)-Aluminum(III) Mixed Oxide. *Ind. Eng. Chem. Res.* **2007**, *46*, 5346–5356.

31. Raichur, A.; Basu, M. Adsorption of Fluoride onto Mixed Rare Earth Oxides. *Sep. Purif. Technol.* **2004**, *24*, 121–127.

32. Vithanage, M.; Jayarathna, L.; Rajapaksha, A.; Dissanayake, C.; Bootharaju, M.; Pradeep, T. Modeling Sorption of Fluoride on to Iron Rich Laterite. *Colloids Surf. A Physicochem. Eng. Asp.* **2012**, *398*, 69–75.

33. Qiao, J.; Cui, Z.; Sun, Y.; Hu, Q.; Guan, X. Simultaneous Removal of Arsenate and Fluoride from Water by Al-Fe (Hydr) Oxides. *Front. Environ. Sci. Eng.* **2014**, *8*, 169–179.

34. Ghorai, S.; Pant, K. Equilibrium, Kinetics and Breakthrough Studies for Adsorption of Fluoride on Activated Alumina. *Sep. Purif. Technol.* **2005**, *42*, 265–271.

35. Kamble, S.; Deshpande, G.; Barve, P.; Rayalu, S.; Labhsetwar, N.; Malyshew, A.; Kulkarni, B. Adsorption of Fluoride from Aqueous Solution by Alumina of Alkoxide Nature: Batch and Continuous Operation. *Desalination.* **2010**, *264*, 15–23.

36. Bouguerraa, W.; Ben Sik Alib, M.; Hamrounia, B.; Dhahbib, M. Equilibrium and Kinetic Studies of Adsorption of Silica onto Activated Alumina. *Desalination.* **2007**, *206*, 141–146.

37. Fan, X.; Parker, D.J.; Smith, M.D. Adsorption Kinetics of Fluoride on Low Cost Materials. *Water Res.* **2003**, *37*, 4929–4937.

38. Lv, L.; He, J.; Wei, M.; Duan, X. Kinetic Studies on Fluoride Removal by Calcined Layered Double Hydroxides. *Ind. Eng. Chem. Res.* **2006**, *45*, 8623–8628.

39. Jagtap, S.; Thakre, D.; Wanjari, S.; Kamble, S.; Labhsetwar, N.; Rayalu, S. New Modified Chitosan-Based Adsorbent for Defluoridation of Water. *J. Colloid Interface Sci.* **2009**, *332*, 280–290.

CHAPTER 7

ADSORPTION OF HEXAVALENT CHROMIUM BY USING SWEET LIME AND ORANGE PEEL POWDER

N. M. RANE[1,*], S. P. SHEWALE[1], S. V. ADMANE[1], and R. S. SAPKAL[2]

[1]Department of Chemical Engineering, MIT Academy of Engineering, Alandi (D), Pune, Maharashtra, India

[2]UDCT, Sant Gadge Baba Amravati University, Amravati, Maharashtra, India

*Corresponding author. E-mail: nitinmrane@rediffmail.com, nmrane@chem.maepune.ac.in

CONTENTS

ABSTRACT

Over the last century, continued population growth and industrialization has resulted in the degradation of a variety of water ecosystems. Such pollution is primarily caused by the discharge of inadequately treated industrial and municipal wastewater. Hexavalent chromium is one such toxic heavy metal which is considered to be a major pollutant in wastewater. The critical parameters affecting the adsorption of Cr(VI) are investigated by using both sweet lime and orange peel powder. The adsorption of Cr(VI) is totally depend on parameters like pH, contact time, temperature, and initial adsorbate concentration. It was observed that the percentage of adsorption increased with the increase of agitation speed up to 180 rpm. The percentage of adsorption increased with increase in particle size from 75 to 180, whereas the uptake also increased. The increase in percentage of adsorption and uptake with increase in particle size might be attributed to the increase in specific surface area of the adsorbent. As the pH increased, a sharp decrease in percentage of adsorption was observed and maximum adsorption capacity is observed at pH 1.25.

The adsorption kinetic models are important in the process of removal of toxic heavy metals from the aquatic environment. The equilibrium adsorption isotherms are also analyzed by studying Langmuir and Freundlich adsorption isotherms. The Kinetic modeling is also done by pseudo first order and pseudo second order kinetic models. The different statistical errors such as sum of the square of the error (SSE), sum of the absolute error (SAE), and average relative error (ARE) are calculated. From error analysis and correlation coefficient (R^2), the best fit for kinetic model was observed for pseudo second order kinetic model. The scanning electron microscopy (SEM) and Fourier transform infrared spectroscopy (FTIR) analysis are also done to study the physical parameters of adsorbents. From the result, it can be concluded that sweet lime peel powder is best adsorbent for removal of chromium from wastewater as compared to orange peel powder.

7.1 INTRODUCTION

Continued population growth and industrialization have resulted in the degradation of water ecosystem. Such pollution is caused due to untreated industrial and municipal wastewater. For every living being, water is one of the most important things needed for its survival. Water is required for agriculture, commercial, domestic, and industrial activities which showed contribution in polluting the water sources.[1] With the increase in population growth, water requirement increases and improper waste disposal creates a problem for water resources. The methods used for treating water resources are very expensive, so innovative technologies should have advantages like low cost, require low maintenance, and it should be energy efficient.[2]

Chromium has wide applications in the field of electroplating, leather tanning, dye cement, and photography industries. The effluents from these industries contained large quantities of heavy metals. Trivalent chromium and hexavalent chromium, both isotopes of chromium, are found in wastewater. But hexavalent chromium is more toxic than trivalent chromium because of its carcinogenic and mutagenic effect. Hence, removal of Cr(VI) from wastewater is the problem faced by the world in today's technology.[3]

Before discharging wastewater containing Cr(VI) into the environment, it should be treated by any available methods such as chemical, biological, and adsorption method.[4–7] Because of ease of operation, low cost, and economic feasibility, the adsorption techniques has more advantages than other techniques.

In this batch studies, sweet lime and orange peel powder were utilized in lab scale experiments for the removal of Cr(VI) using synthetic chromium-contaminated wastewater. The parameters namely initial Cr(VI) concentration, adsorbent dosage, and particle size varied for experimentation. It was found that orange peel powder was to be much effective for removal of Cr(VI) from wastewater.

7.2 MATERIALS AND METHODS

7.2.1 MATERIALS

The adsorbent sweet lime peel (A) and orange peel (B) powder used in this study were collected from the nearby area of Alandi, Pune, Maharashtra. The material was washed with water and dried in sunlight. The dried material was ground into fine powder. Then grinded material was separated in sieve shaker into four different particle sizes.

7.2.2 ADSORBENT CHARACTERIZATION

7.2.2.1 SEM IMAGES FOR ADSORBENT A AND B BEFORE AND AFTER ADSORPTION

A comparison of the scanning electron microscopy (SEM) images before (Figs. 7.1 and 7.3) and after (Figs. 7.2 and 7.4) adsorption showed change of morphology and increase in grain size due to adsorption on the surface for adsorbent A and B. The increase of Cr count after adsorption on adsorbent A and B unequivocally proved the adsorption of the said ion on the adsorbent.

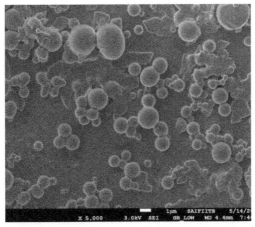

FIGURE 7.1   Before adsorption (A).

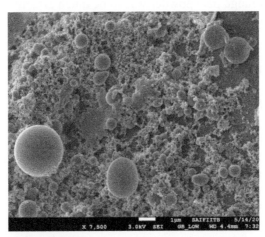

FIGURE 7.2   After adsorption (A).

FIGURE 7.3 Before adsorption (B).

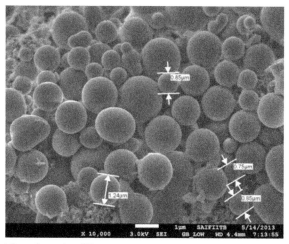

FIGURE 7.4 After adsorption (B).

7.2.2.2 FTIR IMAGES FOR ADSORBENT A AND B BEFORE AND AFTER ADSORPTION

Complexes in the adsorbents are formed during the adsorption apart from electrostatic forces of attraction. The peaks such as 3450 cm^{-1} OH group, 2920 cm^{-1} for aliphatic C—H group, 1670 cm^{-1} for C=O stretching, and 1035 cm^{-1} for C—C stretching by adsorbent is shown in the Figures 7.5–7.8 for before and after adsorption of A and B, respectively. This graph clearly shows the shift or variation in peaks of adsorption. This is due to the major role played by the functional group in the process.

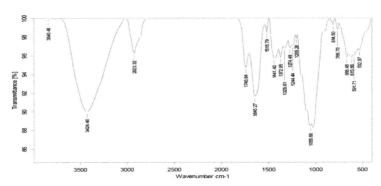

FIGURE 7.5   FTIR before adsorption (A).

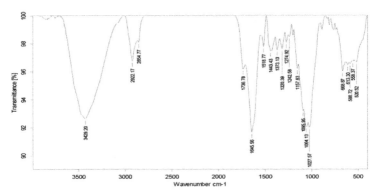

FIGURE 7.6   FTIR after adsorption (A).

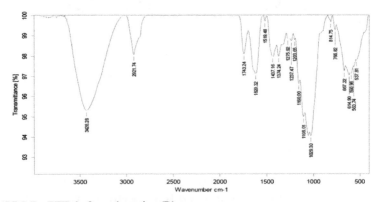

FIGURE 7.7   FTIR before adsorption (B).

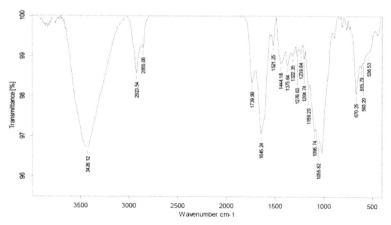

FIGURE 7.8 FTIR after adsorption (B).

7.2.3 ADSORPTION STUDY

The experiments were carried by preparing the stock Cr(VI) solution of 1000 mg/L in distilled water by preparing known amount of analytical grade potassium dichromate powder. The pH of the solution was maintained by adding known concentration of NaOH and HCl. The experiment were conducted for different particle mesh sizes (>75, 75, 150, and 180), solid loading (1, 2, 3, 4, and 5 g) per 250 mL of Cr(VI) solution, and different concentrations (100, 200. 300, 400, and 500 ppm) of the solution. The UV–Vis spectrophotometer (Thermo Fischer 840-210800) is used to analyze the sample. The experiments were continuously done for the required time period of 6 h, pH at 1.25, and agitation speed 180 rpm. In this study, the effect of various parameters on the adsorption of Cr(VI) was studied, and the obtained data were optimized by using pseudo second order kinetic model equation.

7.3 RESULTS AND DISCUSSIONS

7.3.1 EFFECT OF PARTICLE SIZE

The studies to evaluate the optimum particle size were conducted by varying the particle mesh size from 75 to 180 to determine the maximum Cr(VI) removal. The obtained results were shown in Figures 7.9 and 7.10 for percentage removal and uptake, respectively. This might be attributed to increase in specific surface area of the adsorbent.

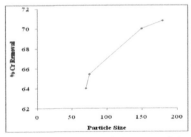
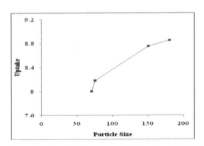

FIGURE 7.9   Uptake and Cr removal for adsorbent A for different particle size (>75, 75, 150, and 180).

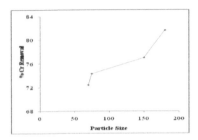
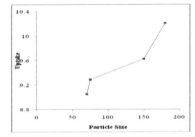

FIGURE 7.10   Uptake and Cr removal for adsorbent B for particle size (>75, 75, 150, and 180).

7.3.2   EFFECT OF INITIAL CONCENTRATION OF Cr(VI)

The percentage removal of Cr(VI) with different adsorbate concentrations were studied. The concentrations were varied from 100 to 500 ppm, keeping adsorbent dose at (2 g/250 mL), stirring speed (180 rpm), pH (1.25), and particle mesh size 180. These results are shown in Figures 7.11 and 7.12, respectively. From the graph, it can be concluded that with the increase in adsorbate concentration, percentage adsorption decreases. While the reverse trend can be observed for Cr(VI) uptake capacity.

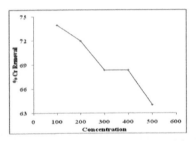
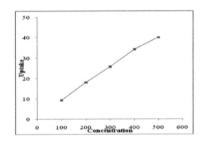

FIGURE 7.11   Uptake and Cr removal for adsorbent A for different initial concentration.

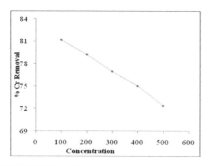
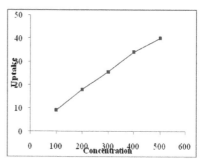

FIGURE 7.12   Uptake and Cr removal for adsorbent B for different initial concentration.

7.3.3   EFFECT OF ADSORBENT DOSE

The effect of adsorbent dose for the adsorption of Cr(VI) from aqueous solution were studied by varying adsorbent doses at various time intervals at a condition of initial Cr(VI) concentration of 100 ppm (Figs. 7.13 and 7.14). Increasing adsorbate concentration increases higher percentage of adsorption because it has more binding sites of adsorption as well as more surface area.[8] The increase in adsorbent dose decreases uptake because of overlapping of adsorption sites due to overcrowding of adsorbent particles.

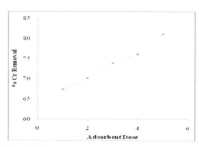
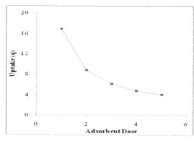

FIGURE 7.13   Uptake and Cr removal for adsorbent A for different adsorbent dosage.

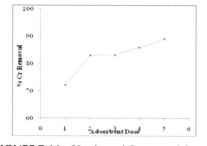
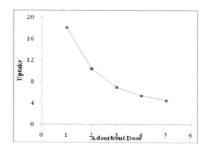

FIGURE 7.14   Uptake and Cr removal for adsorbent B for different adsorbent dose.

7.3.4   KINETIC MODEL

The kinetic modeling study was done for the effect of initial and final concentration at different time intervals. The driving force is the difference in concentration of the average solid phase and equilibrium in adsorption.

7.3.4.1   PSEUDO FIRST ORDER

The sorption of solute from a liquid solution is expressed by using pseudo first order equation. The equation is expressed as:

$$\frac{dq_e}{dt} = k(q_e - q) \qquad (7.1)$$

where

q_e—the amount of solute adsorbed at equilibrium per unit mass of adsorbent (mg/g),

q—the amount of solute adsorbed at any given time "t,"

k—rate constant.

7.3.4.2   PSEUDO SECOND ORDER

The kinetics of adsorption is also explained by pseudo second order. The equation is expressed as:

$$\frac{dq}{dt} = k\left(q_e - q\right)^2 \qquad (7.2)$$

where

q—the amount of solute adsorbed at any given time "t,"

q_e—the amount of solute adsorbed at equilibrium per unit mass of adsorbent (mg/g).

Equation 7.2 can be integrated by applying conditions to obtain a linear relationship between ratio of t/q vs. "t." Different adsorption parameters will give a linear relationship, which allows for computation of "q_e" and "k" (refer Table 7.1).

7.3.5 EQUATION OF ERROR ANALYSIS

Error analysis is used to find best model for fitting experimental value with analyzed data by using the sum of the square of the error (SSE), the sum of the absolute error (SAE), average relative error (ARE), and the average relative standard error (ARS). The equations are expressed as:[9–12]

$$SSE = \sum_{i=1} (y_c - y_e)^2_i \qquad (7.3)$$

$$SAE = \sum_{i=1} |y_c - y_e|_i \qquad (7.4)$$

$$ARE = \frac{1}{n} \sum_{i=1}^{n} \frac{|y_C - y_e|}{y_e} \qquad (7.5)$$

$$ARS = \sqrt{\frac{\sum_{i=1}^{n} \left[\frac{y_C - y_e}{y_e}\right]^2}{n-1}} \qquad (7.6)$$

TABLE 7.1 Kinetic Model Parameters and Error Analysis for Pseudo First and Second Order for Orange Peel Powder (B).

Parameters	Order	R^2	k	q_eCal	q_eExp	SSE	SAE	ARE	ARS
Particle size	Pseudo 2nd	0.999	0.0091	9.828	9.543	0.092	0.286	0.0295	0.0007
	Pseudo 1st	0.486	0.0005	1.366	9.543	67.021	8.176	0.856	0.734
Adsorbent dosage	Pseudo 2nd	0.997	0.012	9.273	9.033	0.069	0.048	0.03	0.006
	Pseudo 1st	0.686	0.0083	1.610	9.033	75.737	7.423	0.795	0.635
Initial Cr(VI) concentration	Pseudo 2nd	0.998	0.005	29.261	28.320	1.439	0.941	0.03	0.0009
	Pseudo 1st	0.719	0.010	2.448	28.32	810.816	28.876	0.899	0.811

7.3.6 ISOTHERM STUDY

The desorption mechanism between adsorbate and adsorbent molecules, efficiency of industrial adsorbate, and use of different types of adsorbents are studied by using isotherms curves.

7.3.6.1 LANGMUIR ISOTHERM

In sorbate–sorbent system, for describing equilibrium conditions for sorption behavior, Langmuir isotherms are used. The equations can be written as:

$$\frac{c_e}{q_e} = \frac{1}{Q_o.b} + \frac{c_e}{Q_o} \tag{7.7}$$

where c_e—the equilibrium concentration,

q_e—the amount of adsorbate adsorbed per gram of adsorbent at equilibrium (mg/g),

Q_o and b—Langmuir constants related to the sorption capacity and intensity, respectively.

7.3.6.2 FREUNDLICH ISOTHERM

The Freundlich adsorption equation can be written as:

$$\frac{x}{m} = q_e = k_f . c_e^{\frac{1}{n}} \tag{7.8}$$

where q_e—equilibrium adsorption capacity (mg/g),

c_e—the equilibrium concentration of the adsorbate in solution,

k_f and n—constants related to the adsorption process such as adsorption capacity and intensity, respectively.

The Freundlich and Langmuir constant are reported in Table 7.2.

TABLE 7.2 Freundlich and Langmuir Constants.

Parameter	Adsorbent	Freundlich isotherm constant			Langmuir isotherm constant		
		n	K_f	R^2	Q_o	b	R^2
Particle size	A	−2.074	43.641	0.999	0.176	0.570	0.999
	B	3.44	23.571	0.994	7.46	−0.21	0.998
Adsorbent dosage	A	0.418	3.549×10^{-3}	0.845	0.245	−0.104	0.779
	B	0.64	0.106	0.917	3.76	0.22	0.776
Adsorbate concentration	A	0.225	7.725×10^{-6}	0.893	5.681	−0.027	0.727
	B	0.26	2.118×10^{-4}	0.947	7.82	−0.03	0.770

7.4 CONCLUSION

In this study, the adsorption parameters such as particle size, initial Cr(VI) concentration, and adsorbent dosage were varied at pH 1.25, agitation speed 180 rpm. With the increase in particle size, both percentage removal and uptake will increase. By increasing adsorbate concentration, percentage of adsorption decreased while exactly reverse trend is observed in uptake capacity. When increasing adsorbent dose from 1 to 5 g, adsorption percentage increases from 70 to 90 while uptake decreases due to overlapping of adsorption sites. From results, it was observed that second order kinetics was followed by adsorption processes. From experimental result, it was observed that experimental data showed good agreement with Langmuir isotherm because of maximum uptake values. The orange peel powder (B) has greater potential as compared to sweet lime peel powder (A).

KEYWORDS

- sweet lime peel
- orange peel
- SEM
- Cr(VI)
- adsorbent dose
- kinetic model

REFERENCES

1. Mukhopadyay, B.; Sundquist, J.; Schmitz, R. J. Removal of Cr(VI) from Cr-Contaminated Groundwater Through Electrochemical Addition of Fe(II). *J. Environ. Manage.* **2007,** *82,* 66–76.
2. Kendelewicz, T.; Liu, P.; Doyle, C. S.; Brown, G. F. Spectroscopic Study of the Reaction of Aqueous Cr(VI) with Fe_3O_4 (III). *Surf. Sci.* **2000,** *469,* 144–163.
3. Gao, P.; Gu, X.; Zhou, T. New Separation Methods for Chromium (VI) in Water by Collection of its Ternary Complex on an Organic Solvent Soluble Membrane Filter. *Anal. Chem. Acta.* **1996,** *332,* 307–312.
4. Baral, S. S.; Das, S. N.; Rath, P. Hexavalent Chromium Removal from Aqueous Solution by Adsorption on Treated Saw Dust. *Biochem. Eng. J.* **2006,** *31,* 216–222.
5. Chun, L.; Hongzhang, C.; Zuohu, L. M. Adsorptive Removal of Cr (VI) by Fe-Modified Steam Exploded Wheat Straw. *Process Biochem.* **2004,** *39,* 541–545.

6. Baral, S. S.; Das, S. N.; Rath, P.; Chaidhary, G. R. Cr (VI) Removal by Calcined Bauxite. *Biochem. Eng. J.* **2007**, *34*, 69–75.

7. Baral, S. S.; Das, S. N.; Chaudhary, G. R.; Swamy, Y. V.; Rath, P. Adsorption of Cr(VI) Using Thermally Activated Weed *Salvinia cucullata*. *Chem. Eng. J.* **2008**, *139*, 245–255.

8. Garg, V. K.; Gupta, R.; Kumar, R. Adsorption of Chromium from Aqueous Solution on Treated Saw Dust. *Bioresour. Technol.* **2004**, *92*, 78–81.

9. Lazaridis, N. K.; Asouhidou, D. D. Kinetics of Sorptive Removal of Chromium(VI) from Aqueous Solutions by Calcined Mg–Al–CO$_3$ Hydrotalcite. *Water Res.* **2003**, *37*, 2875–2882.

10. Leyva, R. R.; Bernal, J. L. A.; Acosta, R. I. Adsorption of Cadmium (II) from Aqueous Solution on Natural and Oxidized Corncob. *Sep. Purif. Technol.* **2005**, *45*, 41–49.

11. Aksu, Z.; Isoglu, I. A. Removal of Copper (II) Ions from Aqueous Solution by Biosorption onto Agricultural Waste Sugar Beet Pulp. *Process Biochem.* **2005**, *40*, 3031–3044.

12. Kundu, S.; Gupta, A. K. Arsenic Adsorption onto Iron Oxide Coated Cement (IOCC): Regression Analysis of Equilibrium Data with Several Isotherm Models and Their Optimization. *Chem. Eng. J.* **2006**, *122*, 93–106.

CHAPTER 8

INVESTIGATION ON ELIMINATION OF CR(VI) FROM WASTEWATER BY POWDERED SHELL OF PEAS AS ADSORBENT

V. S. WADGAONKAR[1,*] and R. P. UGWEKAR[2]

[1]Department of Petrochemical Engineering, Maharashtra Institute of Technology, Pune411038, Maharashtra, India

[2]Department of Chemical Engineering, Laxminarayan Institute of Technology, Nagpur 440033, Maharashtra, India

[*]Corresponding author. E-mail: vinayak.wadgaonkar@mitpune.edu.in

CONTENTS

ABSTRACT

Due to increase in population coupled with mining, extraction, and use of various metals as different industrial and household materials, the load of toxic metal pollution in the environment is increasing. Toxic metals can be hazardous even at very low concentration. When they get into water supplies and aqueous environments the health of plants and animals as well as humans can be impaired. The demand of chromium has been increasing globally because of its extensive use in various metallurgical, chemical, and leather tanning industries due to its various physico-chemical properties. The adsorption of toxic metals is strongly dependent on pH, temperature, contact time, and initial adsorbate concentration. The studies on adsorption were conducted by varying various parameters such as contact time, amount of adsorbent, temperature, and agitation speed. The increase of adsorbent dose provides more surface area for adsorption hence increased the adsorption capacity of the adsorbent. Agitation speed has little effect on adsorption. Experimental data for the adsorbent was fitted to different isotherm models such as Freundlich and Langmuir isotherm. Langmuir adsorption isotherm was employed in order to evaluate the optimum adsorption capacity of the adsorbent. It was concluded that adsorbent prepared from Peas Shell can be used for removal of Cr(VI) from wastewater.

8.1 INTRODUCTION

Agricultural, industrial, domestic, and commercial sector need acceptable-quality water and human life requires high-quality water. Water is polluted by all these activities. Fresh water bodies every day are polluted by billions of gallons of waste from all these resources. Due to improper waste disposal, slowly all water resources are becoming unusable while the requirement for water is increasing. Innovative technologies are having great demand because of small price, need little repairs, as well as are energy proficient. So technique of giving better treatment on behalf of every polluting source is hard and also costly. Chromium is used in leather tanning, metallurgy, and electroplating along with other industries due to which its concentration is elevated in marine systems. Chromium and its salts are used in electroplating, alloying, leather tanning, and rust guard industries;[1] as a result, their requirement is rising through instance. Waste along with waste matter streams of chromite mining and processing contains chromium having two oxidation states, Cr(VI) and Cr(III). Since Cr(VI) type is movable, carcinogenicity is due to oxidizing characters.[2] Due to toxicity, Cr(VI) is considered because the suggested boundary of Cr(VI) inside drinkable water is not more than 0.05 mg/L.[3] Manufacturing as well as mining waste matter contains elevated concentrations of Cr(VI) which is higher than permitted maximum value. For keeping clean environment, treatment of effluent is needed to reduce pollutant. The reduction of pollution due to Cr(VI) consists of two main processes: reduction of Cr(VI) to Cr(III) on the way to make it harmless and removal of Cr(VI). Wastewater containing Cr(VI) is treated by reduction followed by precipitation technique. Reduction followed by precipitation technique is having solid–liquid separation along with throwing away of mire key trouble. Adsorption, ion exchange, and membrane separation are alternate methods used for treatment of Cr(VI)-containing wastewater. Problems encountered in the membrane system are scaling, fouling, and jamming. Ion exchange process is unaffordable because of high price of viable ion exchange resins. Adsorption practice is attractive due to low cost in addition to being precisely easy owing to minimum necessity of management method. In recent times, treatment of Cr(VI)-containing wastewater is done by a number of inexpensive non-conventional adsorbents.[4–8]

8.2 MATERIALS AND METHODS

Peas (seed cover) aggregated from grassland, dirt, and other particulate matter were removed by washing in running tap water. Then the shell was oven dehydrated at 80°C for 24 h. The peas shell was made into fine particles. Once dehydrated at 80°C, it was crushed as well as screened (by means of screen through mesh size 100). Plastic-stopper bottles (containers) were used to conserve the powder in order to decrease contact with moisture; all bottles were conserved in desiccators prior to the time of utilize.

Experiments were carried out to conclude the competence of peas shell in metal adsorption via diverse amounts of adsorbent, that is, 0.5, 1, 1.5, 2, 2.5, and 3 g into particular solution. Runs were performed by altering contact period and agitation speed.

Conical flask of 250 mL was used to carry out experiment and 125 mL was the total volume of reaction mixture. The filling of flasks were cleaned all the way with Whitman No. 40 filter paper after corresponding contact time to keep away from the possible intervention of turbidity. UV–visible spectrophotometry was used to determine chromium concentration of filtered solutions. Adsorbate and the adsorbent contact time are of great consequence in the wastewater handling as a result of adsorption.

8.3   ADSORPTION KINETICS

Pseudo first-order kinetics was used to explain the kinetics of removal of Cr(VI). Adsorption kinetics of heavy metal ions was explained by carrying out experiments in support of contact period varying as of 1–360 min studying elimination of Cr(VI).

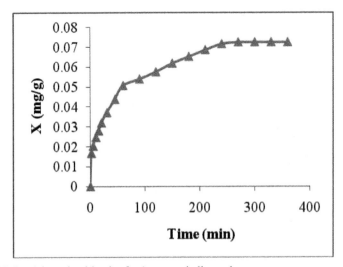

FIGURE 8.1   Adsorption kinetics for 1 g peas shell powder.

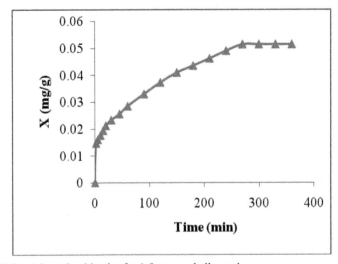

FIGURE 8.2   Adsorption kinetics for 1.5 g peas shell powder.

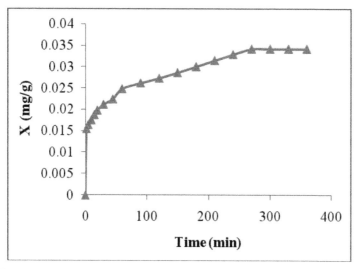

FIGURE 8.3   Adsorption kinetics for 2.5 g peas shell powder.

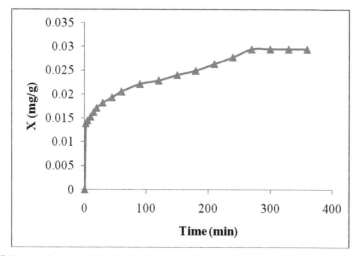

FIGURE 8.4   Adsorption kinetics for 3 g peas shell powder.

Data on the kinetics of solute uptake are mandatory, intended for choosing best working conditions for complete batch method. Figures 8.1–8.4 show graph for the metal uptake: X (mg/g) against time t (min) for early solute concentration of 100 mg/L. From the figure, it can be recognized that X value enlarged with boost in contact time. Different doses of adsorbent X value change with time in the given process.

8.4   RESULTS AND DISCUSSIONS

Contact period has important consequence on percentage adsorption. Percentage adsorption against early concentration at varying contact period was investigated. It could be understood that percentage adsorption lessened with boost in early concentration of the adsorbate. Conversely, binding capability enhanced with enhancement in early concentration, possibly owing to the elevated figure of Cr(VI) ions inside the solution meant for adsorption. To tackle every mass transfer resistances of the metal ions, higher initial adsorbate concentration provided superior powerful force. This elevated the probability of collision of Cr(VI) ions along with energetic localities. This resulted in the enhancement of uptake of Cr(VI) for essential quantity of modified adsorbent.[9]

8.4.1   EFFECT OF CONTACT TIME

A plot of percentage adsorption against adsorption time is given in Figure 8.5. Linear behavior of adsorption with time is given in the figure. Past 1 h, percentage Cr(VI) elimination is 41.40, subsequently, there is decline in

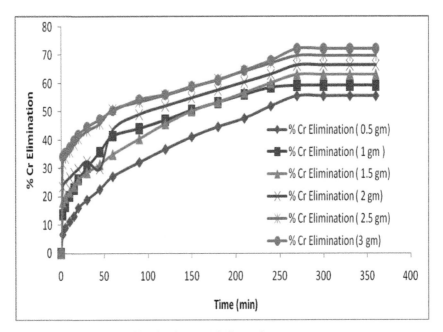

FIGURE 8.5   Adsorption kinetics for peas shell powder.

the chromium elimination due to uninterrupted adsorption and desorption progression. Smooth curves in the graph indicate the development of a single coat on the surface of the adsorbent. During the premature phase, slope of the graph was one, which lessened with time. It shows that the speed of adsorption was elevated during early phases but unhurriedly abridged as well as became stable after equilibrium was attained.

8.4.2 EFFECT OF CHROMIUM CONCENTRATION ON ADSORPTION PROCESS

A physio-chemical feature of sorption practice gives crucial data intended for estimation of adsorption practice as a chief process. The adsorption isotherm is studied via altering the early concentration of Cr(VI).

8.4.2.1 LANGMUIR ISOTHERM FOR ADSORPTION FOR PEAS SHELL POWDER

The Langmuir adsorption isotherm (Fig. 8.6) has been applied to adsorption processes and has been extensively used for the adsorption of a solute

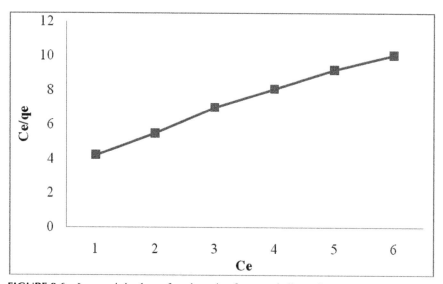

FIGURE 8.6 Langmuir isotherm for adsorption for peas shell powder

from a liquid solution. A fundamental postulation of the Langmuir theory is that adsorption takes place at definite homogeneous localities inside the adsorbent.

The Langmuir equation used is

$$\frac{C_e}{q_e} = \frac{1}{bq_{max}} + \frac{1}{q_{max}C_e}$$

(8.1)

where C_e is the equilibrium concentration (mg/L), q_e the quantity of metal ion sorbed (mg/g); q_{max} the maximum quantity of metal ion sorbed (mg/g) and b is the sorption equilibrium constant (L/mg). A graph of C_e/q_e versus C_e should designate a straight line of slope $1/q_{max}$ and an intercept of $1/bq_{max}$.

8.4.3   EFFECT OF TEMPERATURE

Adsorption practice is influenced by temperature factor. The Cr(VI) adsorption was considered as a function of temperature parameter. Decline in percentage adsorption with increase in temperature is due to desorption caused by the enhancement in the existing thermal energy. Desorption was caused by elevated temperature which promotes elevated movement of the adsorbate. Subsequently, at low pH, acid chromate ions (HCrO$_4^-$) are the foremost variety as revealed by the steadiness figure of Cr(VI)–H$_2$O arrangement.

8.4.4   EFFECT OF AGITATION SPEED

Adsorption runs were performed via early Cr(VI) concentration of 100 mg/L. Agitation speed was altered as of 80–160 rpm. It shows that shakeup rate has small consequence on adsorption practice. It was shown that the percentage of adsorption improved with the enhancement of agitation speed up to 160 rpm (Fig. 8.7) and thereafter, the adsorption competence was independent of agitation speed. The enhancement of adsorption efficiency with the enhancement of agitation speed could be principally due to resistance to mass convey in the bulk solution at lesser shakeup speeds.

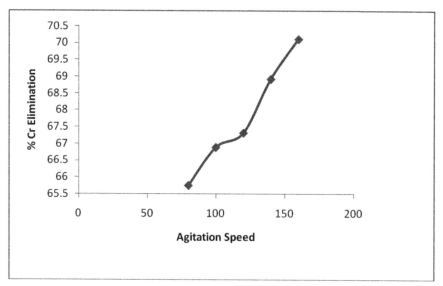

FIGURE 8.7   Agitation speed vs. percentage Cr elimination.

8.5   CONCLUSION

Analyses of consequences of this work have revealed that peas shell is capable of elimination of Cr(VI) from aqueous solutions. The concentration of heavy metal also has a considerable effect under the consequence of this handling. Peas shell is a low-priced matter; subsequently, its use inside industrial wastewater treatment plants would be suitable. Decline in percentage of adsorption with enhancement in temperature shows the technique is exothermic in character and accordingly that lesser temperatures support the adsorption practice.

KEYWORDS

- **chromium (VI)**
- **adsorption isotherms**
- **adsorbent**
- **water pollution**

REFERENCES

1. Erdam, M.; Altundagan, H. S.; Tumen, F. Removal of Hexavalent Chromium by Using Heat Activated Bauxite. *Min. Eng.* **2004,** *17,* 1045–1052.
2. Sarma, Y. C. Cr (VI) from Industrial Effluents by Adsorption on an Indigenous Low Cost Material. *Colloid Surf. A. Phyiochem. Eng. Aspects.* **2003,** *215,* 155–162.
3. Selvaraj, K.; Manonmani, S.; Pattabhi, S. Removal of Hexavalent Chromium Using Distillery Sludge. *Bioresour. Technol.* **2003,** *89,* 207–211.
4. Pradhan, J.; Das, S. N.; Thakur, R. S. Adsorption of Hexavalent Chromium from Aqueous Solution by Using Activated Red Mud. *J. Colloid Interf. Sci.* **1999,** *217,* 137–141.
5. Das, D. D.; Mohapatra, R.; Pradhan, J.; Das, S. N.; Thakur, R. S. Adsorption of Cr(VI) from Aqueous Solution Using Activated Cow Dung Carbon. *J. Colloid Interf. Sci.* **2000,** *232,* 235–240.
6. Sarin, V., Pant, K. K. Removal of Chromium from Industrial Waste by Using Eucalyptus Bark. *Bioresour. Technol.* **2005,** *97,* 15–20.
7. Singh, K. K.; Rastogi, R.; Hasan, S. H. Removal of Cr (VI) from Wastewater Using Rice Bran. *J. Colloid Interf. Sci.* **2005,** *290,* 61–68.
8. Kobya, M.; Demirbas, E.; Senturk, E.; Ince, M. Adsorption of Heavy Metal Ions from Aqueous Solutions by Activated Carbon Prepared from Apricot Stone. *Bioresour. Technol.* **2000,** *96,* 1518–1521.
9. Baral, S. S.; Das, S. N.; Rath, P. Hexavalent Chromium Removal from Aqueous Solution by Adsorption on Treated Sawdust. *Biochem. Eng. J.* **2006,** *31,* 216–222.

CHAPTER 9

REMOVAL OF CALCIUM AND MAGNESIUM FROM WASTEWATER USING ION EXCHANGE RESINS

V. D. PAKHALE[1] and P. R. GOGATE[2,*]

[1]*Department of Chemical Engineering, MIT Academy of Engineering, Alandi, Pune 412105, Maharashtra, India*

[2]*Department of Chemical Engineering, Institute of Chemical Technology, Mumbai 400019, Maharashtra, India*

Corresponding author. E-mail: pr.gogate@ictmumbai.edu.in

CONTENTS

ABSTRACT

Water is a necessity at all levels including industrial applications and drinking purpose. Water to be used in these applications, especially drinking, has to be soft and free from the hazardous metals. Calcium and magnesium are the main hardness-causing metals typically present in water. Apart from issues related to drinking, presence of these elements in water used in industrial applications results in scale formation in processing equipment, creating issues in operations. This work deals with the use of ion exchange beads to remove calcium and magnesium from hard water using both batch and continuous column operations. Indion 225 Na resin has been used as an adsorbent and effect of different operating parameters have been investigated. It was established that the optimum contact time required for the equilibrium was 44 min whereas the optimum pH for maximum removal was 6. The obtained results for equilibrium study have been fitted into different adsorption isotherm models. The Langmuir adsorption isotherm model was observed as the best-fitted model in this system and Langmuir parameters obtained were q_m as 54 and b as 16 confirming better sorption capacity for the used resin.

9.1 INTRODUCTION

Water forms the most important constituent on the earth related to the survival of human beings. The use of fresh water for industrial purposes is a serious problem nowadays faced by the entire world as it leads to scarcity of drinking water. Adding to the problem is the discharge of the effluent by various industries that is contaminating the natural sources of water. To address these issues, it is most important to develop adequate treatment methods for possible recycle of the wastewater, which can also help to avoid major damage to the ecosystem. The treated wastewater can be reused for the same industrial processing minimizing the load on the natural sources. Some of the natural sources including the ground water are unutilized due to high contents of salts especially calcium and magnesium which are very much hazardous for drinking (continuous intake of hard water can cause various diseases like kidney stone) as well as industrial processing. Use of hard water in the heat transfer equipment gives the formation of significant scale, thereby reducing the heat transfer efficacy. Most of the chemical industries need boilers for steam generation and the boiler feed water should be soft in nature, that is, free of calcium and magnesium ions. Also in the case of nuclear power plants, the process water needs to be soft. Thus, in order to avoid damage to the process equipment, metallic elements such as calcium and magnesium are needed to be removed. Also, the effluent from industry needs to be treated for hardness apart from the reduction in organic and inorganic loading before actual recycle. Overall, due to the widespread issues associated with the hardness both in fresh and recycled water, it is imperative to develop newer and efficient treatment schemes for the treatment.

The hard water can be treated using ion exchange resins. Ion exchange resins are nothing but polymeric material with charged ions having the ability to exchange the ions present on the material with the ions present in the wastewater. Such ability is also seen in various natural systems such as soils and living cells. For water purification, synthetic polymeric resins can be used which can also help in removal of minerals.

Mainly, the resins to be used for treating hard water possess Na^+ or H^+ ions suitable for exchange. Calcium and magnesium present in wastewater have more affinity toward this type of resin as these can be easily replaced by the Na^+ or H^+ ions. When the hard water is passed through such type of resin, Ca^{2+} and Mg^{2+} ions are adsorbed on resin by replacing Na^+ or H^+ ions resulting in the formation of soft water based on the removal of hardness-causing elements.[1–3]

9.2 DETAILS ABOUT CALCIUM AND MAGNESIUM

9.2.1 CALCIUM

The atomic number of calcium (Ca) is 20. Calcium is mostly available in the earth's crust. Reactivity of this metal is less as compared to other basic metals. When hard water is used in pipeline, boilers, and heat exchangers, calcium present in the water can deposit on the wall leading to scale formation. Thus, it is imperative to remove Ca during the water processing.

9.2.2 MAGNESIUM

The atomic number of magnesium (Mg) is 12 and it is present in group IIa of the periodic table. Color of the magnesium is silvery white and the relative density is 1.74.

9.2.3   WATER HARDNESS

The normal soft water becomes hard when there is the presence of calcium and magnesium. Generally, the water hardness is measured based on the ability to precipitate soap. Water hardness is defined as the sum of concentrations of calcium and magnesium ions. Water hardness is determined by the ethylenediaminetetraacetic acid (EDTA) titration method.[4]

9.3   MATERIALS AND METHODS

9.3.1   MATERIALS

Indion 225 Na manufactured by Ion Exchange Ltd, Mumbai, India was obtained from Water Treatment Equipment Pvt. Ltd., Pune. This resin, which is a strongly acidic cation exchange resin containing sulfonic acid group, has been used in the study (Fig. 9.1). The resin is extremely robust and has excellent physiochemical properties. It is supplied in moist condition in hydrogen form.

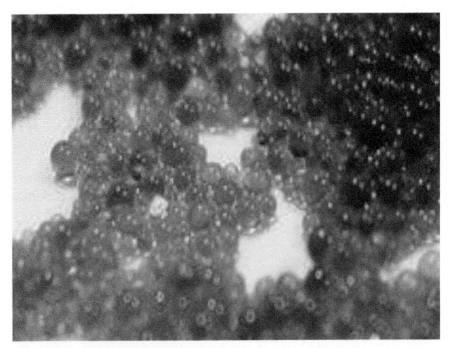

FIGURE 9.1   Indion 225 Na resin.

The other chemicals used in this study were hard water, NaOH, NaCl, EDTA solution, Eriochrome black-T indicator, and distilled water.

9.3.2 METHODS

Initially, the resins were washed with NaOH for conversion into Na^+ form followed by one more wash with distilled water to remove possible organic and inorganic impurities attached to the surface. The resins were then dried in oven at 60°C till constant weight was obtained. The batch experiments were performed in conical flask. A definite amount of dry resin was taken in 50 mL volume of hard water and then subjected to orbital shaking for a predetermined period. The hardness value and percentage metal removal were subsequently determined. The uptake of the metal ions, q_e (mg/g) was decided based on the difference in the original and final concentration. All experiments were performed in duplicate and it was observed that the experimental errors were within 5% of the reported average values.

9.4   RESULTS AND DISCUSSION

9.4.1   EFFECT OF TIME OF CONTACT

The time range selected for the study is 0–56 min and initial metal concentration was 522 mg/L, pH of 6, resin loading as 5 g, and temperature of 25 °C. The effect of the contact time on the hardness elimination is shown in the Figure 9.2.

From the graph, it can be seen that the rate of hardness elimination is more in the initial period and the rate decreases with an increase in the time finally becoming constant after a treatment time of 44 min. The observed results can be attributed to the fact that during the initial time, all the active sites present on the outer layer of the adsorbent are empty and the metal concentration gradient is also more, which means that driving force is significantly higher giving maximum removal of hardness. Further as the time progresses, the driving force reduces and the hardness removal rate decreases with an increase in contact time. From the obtained results, it was established that 44 min is the optimum time for reaching equilibrium and hence was used in the remaining experiments. Similar results have been reported by Swelam et al.[5] where 73% adsorption as equilibrium was observed in 155 min using Resinex K-8H (Na-form) as the adsorbent. It can be also established that the adsorption process is significantly faster in the present work which can be attributed to the better characteristics of the resin used in this work.

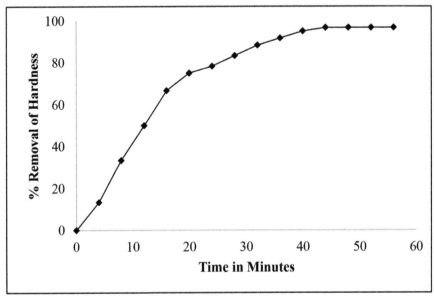

FIGURE 9.2   Effect of contact time on hardness removal.

9.4.2   EFFECT OF PH

The uptake of the metals is affected by the pH of the solution. The effect of pH on adsorption was studied over the pH range of 3–7 at constant operating temperature of 25°C at atmospheric pressure. The obtained results for the effect of pH and percentage removal of calcium and magnesium have been depicted in Figure 9.3. It was observed that the rate of adsorption increases with an increase in pH till an optimum pH of 6. At lower pH value, the adsorption is low due to the more loading of H^+ ions in the solution which opposed the release of fresh H^+ ions from the adsorbent. With an increase in pH, the negatively charged surface area increases which helps in greater metal removal. The maximum adsorption was observed at pH 6, beyond which the extent of adsorption again reduced attributed to the partial hydrolysis of metal ions. Operating pH above 6 gives precipitation of calcium and magnesium ions yielding lower extent of removal. From the graph, it is clearly seen at pH 6, the metal removal was maximum, that is, 84%. Therefore, pH 6 was selected as best possible pH value for the remaining experiments. It is interesting to note that the optimum pH is dependent on the adsorbent–adsorbate system as established by the comparison with reported study on lyocell fibers where maximum calcium-binding capacity of 18–20 mmol/kg was obtained at pH 9.[6]

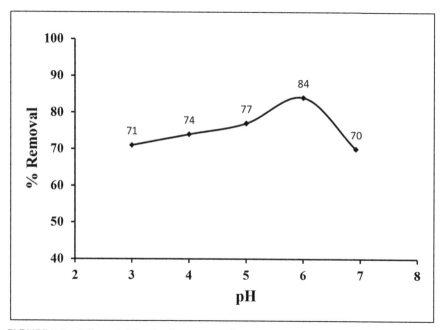

FIGURE 9.3   Effect of pH on hardness removal.

9.4.3   EFFECT OF RESIN LOADING

Effect of resin loading on the degree of Ca^{2+} and Mg^{2+} adsorption was studied at 10 mL/L initial concentration of metals, temperature of 25°C, and pH of 6. The obtained results for the effect of resin loading on the extent of removal of Ca^{2+} and Mg^{2+} have been given in the Figure 9.4. It can be clearly seen that the metal uptake increased with an increase in the adsorbent dose over the entire range of 1–5 mL/L, which can be attributed to the increased adsorbent surface area and availability of higher quantum of adsorption sites. Swelam et al.[5] reported that the adsorption capacity increased till an optimum loading and subsequently reduced at higher dosage due to over-lapping of the adsorption sites attributed to possible agglomeration giving reduction in the total adsorption sites available to metal ions. It is important to understand that the optimum loading was not observed in the present work attributed to the range of resin loading investigated in the work and mixing characteristics. The optimum loading will be dependent on the extent of mixing as well as the type of resin and needs to be established using laboratory scale studies.

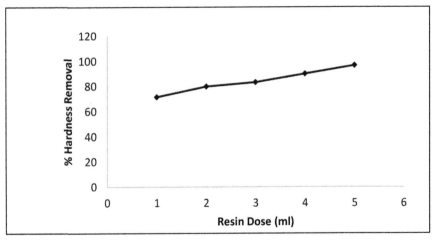

FIGURE 9.4   Effect of resin loading on hardness removal.

9.4.4   EFFECT OF INITIAL METAL ION CONCENTRATION

The initial concentration of calcium and magnesium is an important factor deciding the extent of hardness removal as well as design of a proper treat-ment system. Removal of calcium and magnesium was studied by varying initial concentration over the range of 10–70 mg/L at constant contact time of 44 min and pH of 6. The obtained results have been depicted in the Figure 9.5. It was observed that the extent of hardness removal is marginally affected over the initial range of concentration and mostly constant removal is obtained over the range of 30–50 mg/L. Beyond 50 mg/L loading, lower adsorption was observed. Thus, 40 mg/L as the calcium and magnesium concentration was preferred as the optimized value.

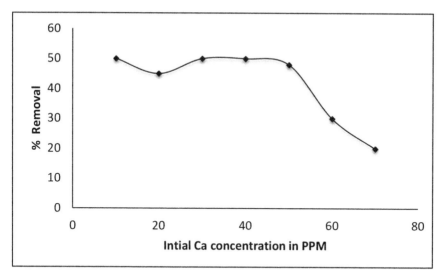

FIGURE 9.5   Effect of initial metal concentration on hardness removal.

9.4.5   ADSORPTION ISOTHERMS

To establish the design equations for adsorption of calcium and magnesium on the Indion 225 Na resin, three isotherm models as Langmuir, Freundlich, and Temkin were selected. Adsorption isotherm plots were drawn with different initial concentrations of calcium and magnesium over the range of 10–50 mg/L keeping the resin amount at a constant level.

9.4.5.1   LANGMUIR ADSORPTION ISOTHERM MODEL

This is possibly the most recognized of all isotherms explaining the adsorption and is given by the following equation:

$$\frac{1}{q} = \frac{1}{q_m bC} + \frac{1}{q_m} \tag{9.1}$$

where "C" is the concentration at equilibrium, "q" is the quantity of Ca and Mg adsorbed per gram of resin at equilibrium condition (mg/g). Constants "q_m" and "b" are associated with sorption capacity and intensity, correspondingly. Langmuir isotherm plot was made for the adsorption of Ca and Mg on Indion 225 Na Resin (Fig. 9.6). From the graph, the obtained value of the constants were $q_m = 54$ mg/g and $b = 16$ L/mg. It can be seen that the sorption capacity and the intensity of the resin are very good confirming good characteristics for the resin and also the fitting of the model was very good as confirmed by R^2 value.

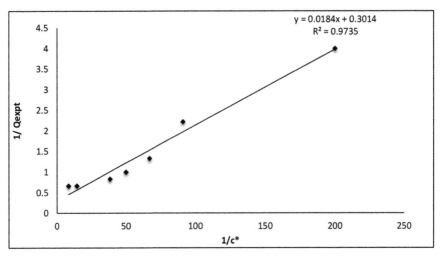

FIGURE 9.6   Langmuir isotherm plot.

9.4.5.2   FREUNDLICH ADSORPTION ISOTHERM MODEL

This equation can be written as:

$$\frac{x}{m} = q_e = kc^{1/n} \tag{9.2}$$

Above equation can be converted into equation of a straight line which can be used to obtain the model parameters as follows:

$$\log q_e = \log k + \frac{1}{n} \log C_e = K_f + \frac{1}{n} \log C_e \tag{9.3}$$

where "q_e" is adsorption capacity (mg/g), "C_e" is the concentration of Ca and Mg in solution at equilibrium, constants "K_f" and "n" are the model parameters giving sorption capacity and intensity respectively. Freundlich isotherm plot was also made for the adsorption of Ca and Mg on Indion 225

Na Resin (Fig. 9.7). From the graph, the values of the constants obtained were $n = 1.8$ and $K_f = 0.8$ though it was observed that the R^2 value was low indicating not so good fitting of the model.

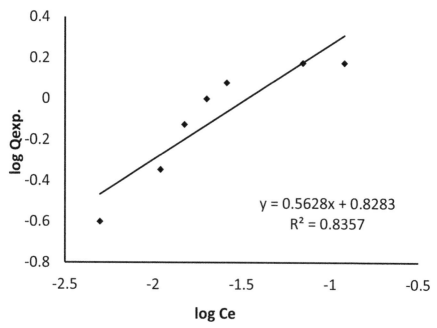

FIGURE 9.7   Freundlich isotherm plot.

9.4.5.3   TEMKIN ADSORPTION ISOTHERM MODEL

Temkin isotherm model assumes that the adsorption heat of all molecules decreases linearly with the increase in coverage of the adsorbent surface. The mathematical form of the equation is as follows:

$$q_{eq} = A + BlogC_{eq} \tag{9.4}$$

where $B = RT/b$, "T" is absolute temperature in Kelvin and C_{eq} is the adsorbate concentration at equilibrium. The constants "R" and "b" are the gas constant and heat of adsorption correspondingly. Temkin isotherm plot (Fig. 9.8) was also plotted for the adsorption of Ca and Mg on Indion 225 Na Resin and the values of the constants obtained were $B = 1$ and $A = 2.6$ mg/g. The R^2 value was found to be lower as compared to the fitting for Langmuir model.

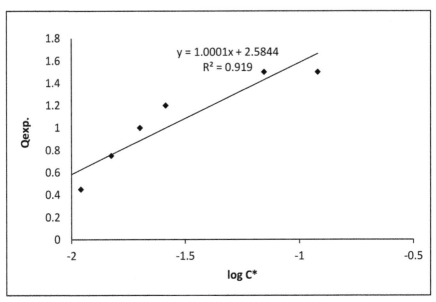

FIGURE 9.8   Temkin isotherm plot.

9.4.6   ADSORPTION KINETIC MODEL

The adsorption kinetics for the hardness removal on the resin was analyzed using two kinetic models as the pseudo first order and pseudo second order model.

9.4.6.1   THE PSEUDO-FIRST-ORDER MODEL

The model can be represented by the following equation,

$$\frac{dq}{dt} = k_1 \ (qe - qt) \tag{9.5}$$

In this equation, k_1 (min^{-1}) is the rate constant for the adsorption, quantity of adsorption at time t (min) is q_t (mg/gm), and quantity of adsorption at equilibrium is given by q_e (mg/g). After integrating and applying boundary conditions, $q_t = 0$ at $t = 0$ and $q_t = q_t$ at time $= t$, above equation can be written in the form of Lagergren-first-order equation[7]

$$\log (q_e - q_t) = \log qe^- (k_1/2.303)t \tag{9.6}$$

The adsorption rate constant k_1, can be calculated by plotting graph of $\log (q_e - q_t)$ versus t (Fig. 9.9). The obtained values of the constants were $k_1 = 0.03$ min^{-1} and $q_e = 0.10$ mg/g with very good fitting of the model. The rate constant values obtained showed a good match with the experimental results reported by Swelam et al.[5]

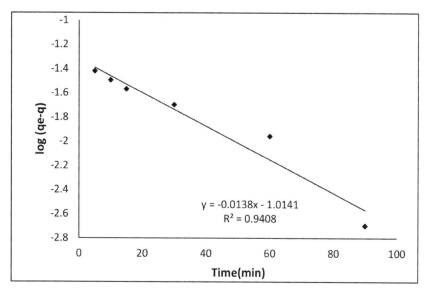

FIGURE 9.9 Pseudo-first-order plot.

9.4.6.2 THE PSEUDO-SECOND-ORDER REPRESENTATION

The model equation is given as,

$$\frac{dq}{dt} = k_2 (q_e - q_t) \tag{9.7}$$

Integration of equation and applying boundary conditions $q_t = 0$ at $t = 0$ and $q_t = q_t$ at $t = t$, we get,

$$\frac{t}{q} = \frac{1}{k2(qe)2} + \frac{t}{qe} \tag{9.8}$$

In this equation, k_2 (g/(mg/min)) is the rate constant value. The values of k_2 and q_e can be known from graph (Fig. 9.10). The obtained values were $k_2 = 0.0005$ (g/(mg/min)) and $q_e = 83.33$ mg/g. For the pseudo-second-order model, the obtained value of correlation coefficient was 0.89, which is marginally lower than that obtained for first order model. Swelam et al.[5] reported that for better fitting of the pseudo-second-order model, the rate decisive step may be chemisorption promoted by covalent forces through the electron replace. In this work, it was observed that the sorption process was better fitting the pseudo-first-order kinetic model with correlation coefficient as 0.94, and standard deviation was also very low. Based on these results, it can be established that sorption of Ca and Mg followed physical adsorption.[5]

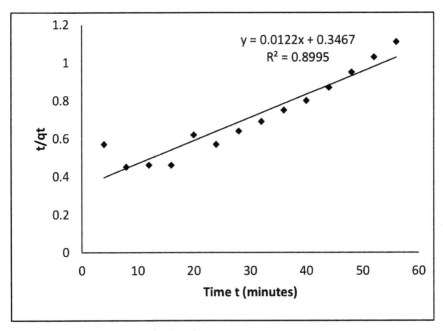

FIGURE 9.10   Pseudo-second-order plot.

9.5   CONCLUSION

The study related to the removal of Ca^{2+} and Mg^{2+} ions using artificial ion exchange resin, Indion 225 Na, demonstrated effective removal of the hardness. The study related to the effect of operating parameters demonstrated that optimum pH was 6 at which maximum adsorption was observed and 44 min was established as the equilibrium time. It was also established that

30–50 mg/L as initial concentration resulted in similar levels of removal. Study of adsorption isotherms revealed that Langmuir adsorption isotherm fitted to the obtained equilibrium data in a better manner than other isotherm models. Overall, the work demonstrated the effective use of ion exchange beads to make soft water from hard water which can be effectively applied for drinking water treatment as well as industrial process water.

KEYWORDS

- ion exchange resins
- hard water
- continuous column operation
- adsorption isotherm

REFERENCES

1. El-Sayed, G. O. Removal of Water Hardness by Adsorption on Peanut Hull. *J. Int. Environ. Appl. Sci.* **2010,** *5,* 47–55.
2. Martinola, F.; Hoffmann, H. Selective Resins and Special Processes for Softening Water and Solutions. *React. Polym. Ion Exch. Sorbents.* **1988,** *7,* 263–272.
3. Bandrabur, B.; Tataru-Farmus, R. E.; Laz, L.; Gutt, G. Application of a Strong Acid Resin as Ion Exchange Material for Water Softening – Equilibrium and Thermodynamic Analysis. *Scientific Study Res. Chem. Chem. Eng. Biotechnol. Food Indus.* **2013,** *13,* 361–370.
4. Rashid, J.; Barakat, M. A.; Alghamdi, M. A. Adsorption of Chromium (VI) from Wastewater by Anion Exchange Resin. *J. Adv. Catal. Sci. Technol.* **2014,** *1,* 26–34.
5. Swelam, A. A.; Salem, A. M. A.; Awad, M. B. Permanent Hard Water Softening Using Cation Exchange Resin in Single and Binary Ion Systems. *World J. Chem.* **2013,** *8,* 1–10.
6. Christa, F. B.; Thomas, B. Sorption of Alkaline Earth Metal Ions Ca^{2+} and Mg^{2+} on Lyocell Fibres. *Carbohydr. Polym.* **2009,** *76,* 123–128.
7. Lagergren, S. Zur Theorie der Sogenannten Adsorption Geloster Stoffe. *K. Sven. Vetenskapsakad Handl.* **1898,** *24,* 1–39.

CHAPTER 10

DEGRADATION OF VEGETABLE OIL REFINERY WASTEWATER USING HYDRODYNAMIC CAVITATION: A PROCESS INTENSIFICATION TECHNIQUE

P. B. DHANKE[1,*] and S. M. WAGH[2]

[1]Chemical Engineering Department, PVPIT, Sangli 416304, Maharashtra, India

[2]Chemical Engineering Department, Laxminarayan Institute of Technology, RTM Nagpur University, Nagpur 440033, Maharashtra, India

*Corresponding author. E-mail: dbpchem@gmail.com

CONTENTS

ABSTRACT

Cavitation using hydrodynamic technique has been proved as an advance treatment option in the treatment of vegetable oil refinery wastewater (VORWW). Properties of VORWW were analyzed in terms of biochemical oxygen demand (BOD), chemical oxygen demand (COD), total organic carbon (TOC), and color. The results given by hydrodynamic cavitation (HC) treatment under controlled conditions were bio-degradability index (BI) of 1.06, COD reduction up to 67.23%, and TOC reduction up to 72.14%, and simultaneously a color reduction up to 55% under the optimum condition of 60 min reaction time. It was found that operating pressure (of 4 bar) in reactor gives the maximum BI (BOD/COD ratio). These results show the power of HC as an effective process intensification technique for the degradation of VORWW.

10.1   INTRODUCTION

Vegetable oil industries are among the highly polluting food industries in the whole world. Vegetable oil refineries generate a large amount of wastewater; in such industries wastewater comes from various operations like degumming, deacidification, deodorization, and neutralization.[1] A number of researchers have used traditional physical-chemical treatment for vegetable oil refinery wastewater (VORWW).[2] The characteristics of VORWW depend on the process implemented for the production of oil, which leads to an increase in chemical oxygen demand (COD), biochemical oxygen demand (BOD), and color. Hence, VORWW has gained importance for environmental protection.

A new frontier technology like cavitation can be efficiently utilized as a treatment option for VORWW. Cavitation is nothing but continuous generation, expansion, and bursting of cavities for a very short time period, which produces large waves of energy on a small area in the reactor system.[3–5] Cavitation produced by varying the inlet pressure of liquid is known as hydrodynamic cavitation (HC) and hence it is highly energy efficient as compared to other methods.[6]

In HC, cavities are generated by passing the wastewater via venture or orifice plates. Then the buildup pressure in vena contracta of this restriction path gets down below the vapor pressure of wastewater, which produces cavities that burst after the orifice plate. The bursting of cavity produces high energy point, which releases highly active free radicals. This leads to high heat, mass, and momentum rates. The collapse of these cavities generates hot points, transferring a temperature ~9000 K and a pressure ~900 atm.[7–14] In this situation, water molecules are fragmented into OH· and H· radicals.

Produced OH· radicals take the path liquid main stream in that they react with pollutants with oxidation reaction. A reason behind the degradation is thermal dissociation of the volatile pollutant which is trapped inside the bubble. While collapsing of cavity, OH· radicals react with the pollutant at the cavity boundary layer. Same things happen with other pollutants. The shockwaves released by the collapsing of cavity can break a high molecular weight compound. Raut-Jadhav et al.[15] used HC with the oxidation technique for the degradation of imidacloprid in aqueous solutions and found that cavitation by hydrodynamic techniques is very useful.

In this research, HC was evolved as a recent treatment option for the VORWW. The VORWW was fed to HC reactor setup. Various parameters were tested during experimentation interns of BOD, COD, and total organic carbon (TOC), which showed an improvement in BOD/COD ratio.

10.2   EXPERIMENTAL METHOD

10.2.1   EXPERIMENTAL SETUP

HC reactor is shown in Figure 10.1. This has a storage (10 L), control valves (V1–V3), the cavity-generating system, and suction-discharge pipe with bypass. The pump has a suction tube from storage and pump discharge is separated into main line (of 25 mm diameter) and bypass. Bypass is used to control the flow of liquid through main tube to the desired pressure. The main tube holds the cavity-generating device which is orifice in this experimentation. Figure 10.2 shows the cavity-generating system (orifice plate) for a setup.

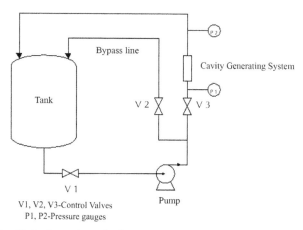

FIGURE 10.1   Hydrodynamic cavitation reactor setup.

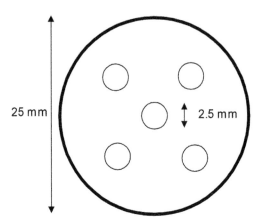

FIGURE 10.2   Schematic of cavity-generating system: an orifice.

10.2.2   EXPERIMENTAL MATERIAL

10.2.2.1   VEGETABLE OIL REFINERY WASTEWATER

VORWW was collected from local oil industry. Wastewater is available in the degumming, deacidification, deodorization, and neutralization process steps. The characteristics of wastewater samples are placed in Table 10.1.

TABLE 10.1   Characteristics of VORWW.

Parameters	Values
BOD	3700
COD	15,260
TOC	14,000
pH	5.1
Color	Yellowish
BI (BOD_5:COD ratio)	0.24

10.2.2.2   REAGENTS AND CHEMICALS

All the required chemicals were purchased as per standard procedure of Central Pollution Control Board (CPCB), India, from local market of AR grade for the determination of BOD, COD, and TOC.

10.2.2.3   ANALYTICAL TECHNIQUES

The BOD, COD, and TOC were determined using analytical procedures of CPCB, India. Total carbon dioxide (due to oxidation) produced during cavitation was counted by measuring TOC. Combustion-infrared method was used to determine TOC. Combustion-infrared instrumental used high temperature, catalysts, and oxygen to convert organic carbon into carbon dioxide (CO_2). Produced CO_2 can be determined by a non-dispersive infrared analyzer. The color determination was also carried out in double beam UV–Visible spectrophotometer.[16]

10.2.2.4   BIO-DEGRADABILITY RATIO

Bio-degradability index (BI) is termed as the ratio of BOD_5:COD. This is the parameter for characterizing bio-degradability of wastewater. The parameter BI gives the sustainability of VORWW for further biological treatment option.[16] BI on the basis of COD and BOD of VORWW was evaluated after HC treatment.

10.2.3   HYDRODYNAMIC CAVITATION TREATMENT ON VORWW

The VORWW was fed to HC treatment in which 5 L of VORWW was degraded in cavitation reactor setup, and cavitation was produced by applying an orifice (as shown in Fig. 10.2). Different experiments were carried out at variable inlet pressures and using different exposure time in reactor. After the specified time interval the samples were taken from the reactor and analyzed for COD, BOD, TOC, pH, and color.

10.3   RESULTS AND DISCUSSION

10.3.1   INLET PRESSURE INFLUENCE ON DEGRADATION OF VORWW

COD/TOC reduction was determined at a minimum (4 bar) and maximum (6 bar) inlet pressure. The inlet COD of VORWW was found to be 15,260 mg/L. The treated samples were collected at 60 min time intervals for the analysis of COD and TOC. Obtained results are given in Table 10.2, which show no drastic decrease in the reduction of COD and TOC with a change in inlet pressure from 4 to 6 bar. On analyzing the results, maximum degradation of VORWW was identified in the initial 60 min of exposure in HC setup. After the 60 min interval, there is no considerable decrease in COD/TOC. This indicates that the optimum range of time was 60 min. The obtained result at pressure of inlet (6 bar) does not indicate further degradation of VORWW. More experiments were carried out at 4 bar pressure, on considering the optimum condition. In this it has been observed that further mineralization rate does not change with varying the pressure of inlet.

10.3.2   HC EFFECT ON COLOR OF VORWW

In the vegetable oil industry, yellowish colored wastewaters mainly come from various routes like degumming, deacidification, deodorization, and neutralization. Using HC technique in this is workable to minimize COD/TOC with a significant reduction in the color of VORWW. At 4 bar inlet pressure in HC, color reduction of 55% was achieved in 60 min. Even after increasing the exposure time up to 120 min, no significant change was found in color reduction. The profile is shown in Figure 10.3.

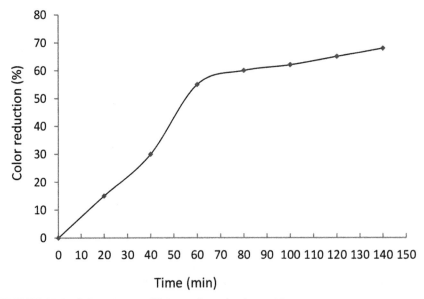

FIGURE 10.3   Inlet pressure effect on color reduction at 4 bar.

The results shown in Figure 10.3 indicate that HC treatment leads to COD/TOC reduction and is even able to reduce toxicity which is shown by the color of wastewater. Maximum of 68% color reduction is found in 120 min and 55% color reduction observed in 60 min time interval which is considered as optimum. The UV–Visible spectrophotometer is used to determine the color of VORWW.

10.3.3 EFFECT OF HYDRODYNAMIC CAVITATION ON BIO-DEGRADABILITY OF VORWW

The BI (BOD/COD ratio) value is nothing but the quantity of waste that can be bio-degraded. In this research work, the degradation treatment of VORWW with HC has increased the values of BI. All the analyzed values are given in Table 10.2. It was noted that there is a drastic change of BI in the first 60 min time interval at 4 bar inlet pressure. On increasing the time up to 120 min, no more change is found in BI. This is a good sign for the operating condition followed in this experiment that gives best BI. At inlet pressure 4 bar BOD enhances significantly. At 6 bar inlet pressure BOD enhancement is not found up to mark. Thus 4 bar inlet pressure has given a better BI as compared to 6 bars. After getting the experimental results, we were able to conclude that the pressure of inlet (4 bar) is favorable in minimizing toxicity (COD/TOC). Similarly for increasing bio-degradability (BOD/COD ratio), an inlet pressure of 4 bar would be preferred. Hence, HC is able to reduce toxicity and similarly increases bio-degradability of VORWW.

TABLE 10.2 Effect of Hydrodynamic Cavitation on Vegetable Oil Refinery Wastewater (VORWW).

Inlet pressure (bar)	Time of exposure (min)	COD (mg/L)	COD removal (%)	TOC (mg/L)	TOC removal (%)	BOD$_5$ (mg/L)	BI (BOD/COD ratio)
4	0	15,260.00	0.00	14,000.00	0.00	3700.00	0.2424
	60	5000.00	67.23	3900.00	72.14	5300.00	1.0600
	120	4910.00	1.80	3810.00	2.30	5310.00	1.0800
6	0	15,200.00	0.00	14,080.00	0.00	3700.00	0.2423
	60	4980.00	67.23	3908.00	72.24	5390.00	1.0823
	120	4910.00	1.40	3870.00	0.97	5400.00	1.0997

Explanation: COD/TOC removal (%) = $(X - D)/(X.100)$, where X = initial COD/TOC (mg/L), D = final COD/TOC (mg/L).

10.4 CONCLUSION

HC is a successful technology for increasing the bio-degradability of VORWW and simultaneously for minimizing toxicity with color. An inlet pressure of 4 bar is the best for minimizing toxicity of VORWW; similarly an inlet pressure of 6 bar has not given more benefit toward minimizing toxicity. Hence, depending on requirements of treatment, HC can be confidently used for the treatment of VORWW.

KEYWORDS

- **hydrodynamic cavitation**
- **bio-degradability**
- **cavity-generating system (orifice plate)**
- **biochemical oxygen demand**

REFERENCES

1. Nweke, C. N.; Nwabanne, J. T.; Igbokwe, P. K. Kinetics of Batch Anaerobic Digestion of Vegetable Oil Wastewater. *J. Water Pollut. Treat.* **2014**, *1*, 2374–6351.
2. Chipasa, K. B. Limits of Physicochemical Treatment of Wastewater in the Vegetable Oil Refining Industry. *Pol. J. Environ. Stud.* **2001**, *3*, 141–147.
3. Jyoti, K. K.; Pandit, A. B. Water Disinfection by Acoustic and Hydrodynamic Cavitation. *Biochem. Eng. J.* **2001**, *7*, 201–212.
4. Vichare, N. P.; Gogate, P. R.; Pandit, A. B. Optimization of Hydrodynamic Cavitation Using a Model Reaction. *Chem. Eng. Tech.* **2000**, *23*, 683–690.
5. Gogate, P. R.; Pandit, A. B. Engineering Design Methods for Cavitation Reactors. II. Hydrodynamic Cavitation. *AIChE J.* **2000**, *46*, 1641–1649.
6. Gogate, P. R.; Pandit, A. B. Hydrodynamic Cavitation Reactors: A State of the Art Review. *Rev. Chem. Eng.* **2001**, *17*, 1–85.
7. Chand, R.; Ince, N. H.; Gogate, P. R.; Bremner, D. H. Phenol Degradation Using 20, 300 and 520 kHz Ultrasonic Reactors with Hydrogen Peroxide, Ozone and Zero Valent Metals. *Sep. Purif. Tech.* **2009**, *67*, 103.
8. Save, S. S.; Pandit, A. B.; Joshi, J. B. Use of Hydrodynamic Cavitation for Large Scale Microbial Cell Disruption. *Chem. Eng. J.* **1994**, *55*, B67.
9. Saharan, V. K.; Badve, M. P.; Pandit, A. B. Degradation of Reactive Red 120 Dye Using Hydrodynamic Cavitation. *Chem. Eng. J.* **2011**, *178*, 100–107.
10. Wang, X.; Zhang, Y. Degradation of Alachlor in Aqueous Solution by Using Hydrodynamic Cavitation. *J. Hazard. Mater.* **2009**, *161*, 202–207.

11. Shah, Y. T.; Pandit, A. B.; Moholkar, V. S. *Cavitation Reaction Engineering;* Plenum Publishers: New York, 1999.

12. Xiaodong, Z.; Yong, F.; Zhiyi, L.; Zongchang, Z. The Collapse Intensity of Cavities and the Concentration of Free Hydroxyl Radical Released in Cavitation Flow. *Chinese J. Chem. Eng.* **2008,** *16,* 547–551.

13. Chakinala, A. G.; Bremner, D. H.; Gogate, P. R.; Namkung, K. C.; Burgess, A. E. Multivariate Analysis of Phenol Mineralisation by Combined Hydrodynamic Cavitation and Heterogeneous Advanced Fenton Processing. *Appl. Catal. B: Environ.* **2008,** *78,* 11–18.

14. Pradhan, A. A.; Gogate, P. R. Removal of P-Nitrophenol Using Hydrodynamic Cavitation and Fenton Chemistry at Pilot Scale Operation. *Chem. Eng. J.* **2010,** *156,* 77–82.

15. Raut-Jadhav, S.; Saharan, V. K.; Pinjari, D. V.; Saini, D. R.; Sonawane, S. H.; Pandit, A. B. Intensification of Degradation of Imidacloprid in Aqueous Solutions by Combination of Hydrodynamic Cavitation with Various Advanced Oxidation Processes (AOPs). *J. Environ. Chem. Eng.* **2013,** *1,* 850–857.

16. Padoley, K. V.; Saharan, V. K.; Mudliara, S. N.; Pandeya, R. A.; Pandit, A. B. Cavitationally Induced Biodegradability Enhancement of a Distillery Wastewater. *J. Hazard. Mater.* **2012,** *219–220,* 69–74.

CHAPTER 11

VISIBLE LIGHT PHOTOCATALYTIC DEGRADATION OF CORALENE DARK RED 2B

M. D. SARDARE[1], P. KOTWAL[1], S. JADHAV [1], S. WAGHELA[1], and S. M. SONTAKKE[2,*]

[1]Department of Chemical Engineering, MIT Academy of Engineering, Alandi, Pune 412105, Maharashtra, India

[2]Department of Chemical Engineering, Institute of Chemical Technology, Mumbai 400019, Maharashtra, India

*Corresponding author. E-mail: sharad.ictmumbai@gmail.com

CONTENTS

ABSTRACT

This study was mainly focused on the sunlight as sources for the photocatalytic degradation of organic dye Coralene Dark Red 2B in synthetic liquid phase. Titanium dioxide was used as catalyst in photocatalytic degradation by means of batch photoreactor. The process of decolorization and removal of organic carbon was effective activity in presence of sunlight. An attempt has been made to study the influence of process parameters namely concentration of dye, dosage of photo catalyst, and speed of magnetic stirrer on photo catalytic degradation of Coralene Dark Red 2B dye. Analysis of the degradation of dye was done by using UV spectrophotometer at wavelength of 549 nm. The appropriate selection of right operating conditions allows effective degradation of dye with sunlight.

11.1   INTRODUCTION

Azo dyes are among the most widely used dyes in textile industry because of their ease of synthesis, versatility, and cost-effectiveness. However, the extensive use of these dye poses a serious threat to the environment because its release is undesirable to the aquatic environment. Most of the commercial azo dyes are chemically stable and are very difficult to remove using conventional wastewater treatment methods,[1] which thereby causes an environmental concern.[2,3] Photocatalytic degradation of azo dyes has received considerable attention in recent years. Photocatalytic degradation of various dyes using different semiconductor catalysts is reported in literature. A process, where sunlight energy equivalent to or more than its band gap energy is absorbed by the semiconductor catalyst, by which the generation of hole and electrons pair takes place, leading to the evolution of free radicals to oxidize the substrate, is called photocatalysis. A wide range of organic compounds are well thought out to be liable for degradation by substantial free radicals.

11.2   MATERIALS AND METHODS

11.2.1   MATERIALS

Analytical grade Coralene Dark Red 2B ($C_{19}H_{17}ClN_5O_7$, 99%) was obtained from Yogesh Dyestuff, Mumbai. Commercial titanium dioxide (TiO_2 (IV)) with a crystallite size of 21 nm was obtained from Hi-Media Chemicals, Mumbai, India. It was used as received. For all experiments only distilled water was used. For the analysis of the degradation samples, a 119 single beam UV–Vis spectrophotometer (Systronics, India) was used.

11.2.2   EXPERIMENTAL METHOD

Photocatalytic degradation of red dye took place in the presence of sunlight having 1125 lux (using Lux meter by VARTECH). Both sides of the setup mirrors were used to increase the intensity of the sunlight. All experiments were performed during 10.30 am to 2.00 pm under direct sunlight. In all photocatalytic degradation experiments, 30 ppm Coralene Dark Red 2B solution of 100 mL was used. A 200 mL beaker (Borosil, India) was used as reactor, which was kept on a magnetic stirrer (Remi, India). To adjust the pH level of the reaction sample, solution of 1 N NaOH/HCl was used. The reaction solution was kept in direct sunlight, with the mirror arranged on both sides of the setup to increase the intensity of light. One milliliter of the degradation sample was taken out for analysis after equal time intervals, filtered to remove the photocatalyst particle and analyzed by UV spectrophotometer at 546 nm. Percentage photocatalytic degradation was calculated from the experimental data and is indicated by %R using the following equation:

$$\% R = \frac{C_0 - C_t}{C_0} \times 100$$

where $\% R$ is the percent photocatalytic degradation of Coralene Dark Red 2B dye; C_0 is the initial concentration of Coralene Dark Red 2B; and C_t is the concentration of Coralene Dark Red 2B at time t.

11.3   MECHANISM OF THE PHOTOCATALYTIC DEGRADATION

Due to photo excitation of photo semiconductor, the photocatalyst of organic matter in the solution was instigated by which an electron hole pair was formed on the surface of the catalyst after the excitation of semiconductor. Nitrogen-to-nitrogen double bond in an azo dye differentiates the mechanism of photocatalytic degradation. Associated chromophores and auxochromes of an azo bond between nitrogen and nitrogen make the color of azo dye resolute. Oxidation of the most active azo bond can be done by positive hole pair or hydroxyl radical reduced by electron in the conduction band. The discoloration of dyes was done because of splitting of $-N=N-$.[4]

Coralene Dark Red 2B dye degradation by photocatalytic mechanism:

$$TiO_2 + h\nu \rightarrow e_{CB}^* + h_{VB}^+ \tag{11.1}$$

$$e_{cv} + O_2 \rightarrow O_2^{*-} \tag{11.2}$$

$$H_2O + O_2^{*-} \rightarrow OOH^* + OH^- \tag{11.3}$$

$$2OOH^* \rightarrow O_2 + H_2O_2 \tag{11.4}$$

$$O_2^{*-} + dye \rightarrow dye - OO^* \tag{11.5}$$

$$OOH^* + H_2O + e_{CB}^* \rightarrow H_2O_2 + OH^- \tag{11.6}$$

$$H_2O_2 + e_{CB}^* \rightarrow OH^* + OH^- \tag{11.7}$$

$$H_2O_2 + O_2^{*-} \rightarrow OH^* + OH^- + O_2 \tag{11.8}$$

$$OH^* \; O_2^{*-} / TiO_2^{++} + dye \rightarrow dye\ degradation \tag{11.9}$$

Here the activity of TiO_2 nanoparticle can be predicted. Due to sunlight irradiation, TiO_2 molecule gets exhilarated and transmits electron to the conduction band, which is shown in eq 11.1. In eq 11.2, super oxide radical is generated from molecular oxygen which diminishes electron in the conduction band of TiO_2. The hole–electron pair recombination process averts molecular oxygen which is adsorbed on the surface of the TiO_2.[5] According to all experiments, the rate of photocatalytic degradation decreases as the recombination of hole–electron pair takes place. From eqs 11.3 to 11.5, radicals may structure hydrogen peroxide or organic peroxide in the occurrence of oxygen and organic molecule. As per the mechanism shown while doing experiments, hydrogen peroxide can be structured as shown in eq 11.6. Hydroxyl radicals are very powerful oxidizing agents, formed from hydrogen peroxide. The produced radicals are very capable to attack dye molecules and degrade them as per eq 11. 9.

11.4 RESULTS AND DISCUSSION

11.4.1 EFFECT OF SUNLIGHT

All experiments were conducted under the sunlight which plays an important role in the degradation of Coralene Dark Red 2B dye. Figure 11.1 shows that the experiments of photocatalytic degradation of Coralene Dark Red 2B dye in distilled water solution carried out below dissimilar experimental conditions, that is, with presence of only sunlight, without sunlight, with only presence of TiO_2, and with both sunlight and TiO_2 as semiconductor photocatalyasts. The percentage degradation increased with the increase in irradiation time, and at 90 min 98% degradation was achieved for Coralene Dark Red 2B dye. At 200 rpm, 30 ppm solution was magnetically stirred with optimum dosage of photocatalyst in sunlight for pH 7, and 72% of degradation was achieved. While exposing dye solution directly to sunlight without adding photocatalyst, degradation was not found because no photo-active catalyst is present, but for with sunlight and catalyst without stirring percentage degradation was 98% within 90 min.

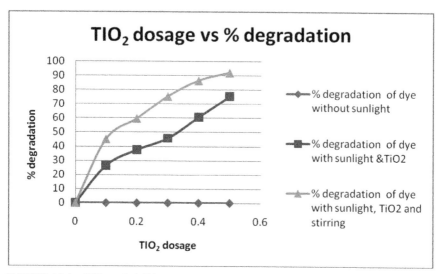

FIGURE 11.1 Effect of visible sunlight on Coralene Dark Red 2B dye by photocatalytic degradation at 30 ppm/100 mL solution and 0.2 TiO_2 in presence of sunlight.

11.4.2   EFFECT OF CATALYST DOSE ON THE RATE OF PERCENTAGE DEGRADATION

For 30 ppm of Coralene Dark Red 2B solution, the semiconductor photo-catalytic dosage varied from 0.1 to 0.5 g at constant pH ~7 levels (Fig. 11.2). A waste solution of 30 ppm was stirred after adding the catalyst to scatter all the way through the solution and kept in sunlight. As this solution was kept in sunlight, degradation started immediately. Average minimum percentage degradation was done at 10 min of 53% at all doses of catalyst. At 0.2 g semiconductor photocatalytic dosage got ~100% at time 90 min. As semiconductor photocatalytic dosage increased in amount, active sites on the photocatalyst surface increased, which resulted in increasing hydroxyl radicals. After some time, it would obstruct the transmission of solar light in the reaction volume and can consequently decrease the percentage decolorization. On the other hand, the decolorization in the initial stages at lower catalyst dosages was faster compared to higher catalyst concentrations due to the decrease in light penetration which hinders the creation of active species.

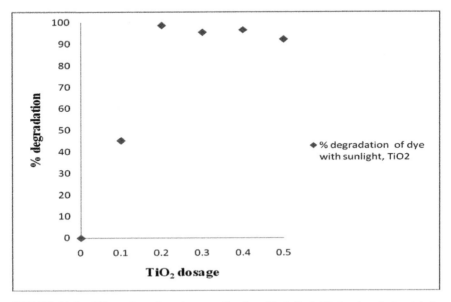

FIGURE 11.2   Effect of catalyst dose on Coralene Dark Red 2B dye by photocatalytic degradation at 30 ppm/100 mL solution in presences of sunlight for 90 min.

11.4.3   EFFECT OF CORALENE DARK RED 2B DYE CONCENTRATION ON THE RATE OF PERCENTAGE DEGRADATION

While studying the effect of Coralene Dark Red 2B dye concentration on the rate of percentage degradation by photocatalytic degradation using TiO_2 semiconductor photocatalyst, it was shown that the rate of percentage degradation diminished with amplification in Coralene Dark Red 2B dye concentration in the presence of solar irradiations. The experiment was done for the range of 10–50 mg/L dye concentration; 0.1 g/100 mL of catalyst dosage at pH 7 for 90 min. Figure 11.3 shows the obtained result. Active sites of catalyst decrease as equilibrium degradation takes place, which results in the formation of primary oxidizing agents OH* radical of azo dye.[6]

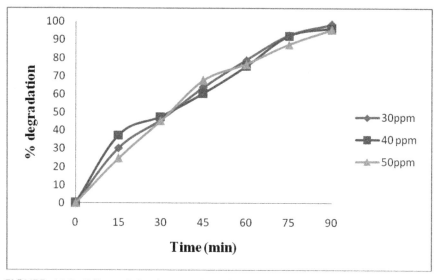

FIGURE 11.3 Effect of Coralene Dark Red 2B dye concentration by photocatalytic degradation at 100 mL solution, in presences sunlight and of 0.2 g TiO_2.

11.5   CONCLUSION

The above study shows that photocatalytic degradation is a very helpful technique for the dilapidation of azo dyes like Coralene Dark Red 2B. In this study, commercial TiO_2 nanoparticles of size 25 nm were used for the degradation of azo dye under the influence of sunlight without stirring. The height of percentage degradation ~100% was found at 30 mg/L dye concentration; 0.2 g/100 mL of catalyst dosage at pH 7 for 90 min. As per the conditions for experiments, it is shown that photocatalytic degradation is cost-effectively practicable compared to other oxidative processes since it rivets sunlight.

KEYWORDS

- sunlight
- TiO$_2$ (titanium dioxide)
- photocatalytic degradation
- Coralene Dark Red 2B
- dye decolorization

REFERENCES

1. Disha Meena, R. C. A Study on Rate of Decolorization of Textile Azo Dye Direct Red 5B by Recent Develop Photocatalyst. *Int. J. Sci. Res. Pub.* **2013,** *3,* 1–4.
2. Salker, A. V.; Gokakakar, S. D. Solar Assisted Photocatalytic Degradation of Amido Black 10B Over Cobalt, Nickel and Zinc Metalloporphyrins. *IJPS.* **2009,** *4,* 377–384.
3. Wu, S.; Zheng, H.; Wu, Y.; Lin, W.; Xu, T.; Guan, M. Hydrothermal Synthesis and Visible Light Photocatalytic Activity Enhancement of BiPO$_4$/Ag$_3$PO$_4$ Composites for Degradation of Typical Dyes. *Ceram. Int.* **2014,** *40,* 14613–14620.
4. Zaman, S.; Zainelabdin, A.; Amin, G.; Nour, O.; Willander, M. Efficient Catalytic Effect of CuO Nanostructures on the Degradation of Organic Dyes. *J. Phys. Chem. Solids.* **2012,** *73,* 1320–1325.
5. Roushania, M.; Mavaeia, M.; Rajabib, H. R. Graphene Quantum Dots as Novel and Green Nano-Materials for the Visible-Light-Driven Photocatalytic Degradation of Cationic Dye. *J. Mol. Catal. A: Chem.* **2015,** *409,* 102–109.
6. Kordouli, E.; Bourikas, K.; Lycourghiotis, A.; Kordulis, C. The Mechanism of Azo-Dyes Adsorption on the Titanium Dioxide Surface and Their Photocatalytic Degradation Over Samples with Various Anatase/Rutile Ratios. *Catal. Today.* **2015,** *252,* 128–135.

PART II
Novel Separation Techniques

CHAPTER 12

MICROWAVE-ASSISTED EXTRACTION OF CARVONE FROM *CARUM CARVI*

O. D. BOROLE[1], S. R. SHIRSATH[1,*], and S. G. GAIKWAD[2]

[1]*Department of Chemical Engineering, Sinhgad College of Engineering, Vadgaon Budruk, Pune 411041, Maharashtra, India*

[2]*Chemical Engineering and Process Development Division, National Chemical Laboratory, Pashan, Pune 411008, Maharashtra, India*

*Corresponding author. E-mail: srshirsath@gmail.com

CONTENTS

ABSTRACT

In this work an attempt was made to find a suitable method for extraction of carvone from caraway seeds (*Carum carvi*). Conventional method of Soxhlet extraction and the non-conventional techniques such as ultrasound-assisted extraction (UAE) and microwave-assisted extraction (MAE) were compared. MAE was found to be a better method as compared with the other two methods. Various parameters such as extraction time, seed to solvent ratio, particle size, and microwave power have been studied using MAE to obtain the optimum values for maximum yield of carvone. Maximum yield of 10.03 mg carvone/g was obtained using Soxhlet extraction after 3 h at 69°C. However, 8.77 mg carvone/g seed was obtained with MAE in just 30 min at conditions of 1:10 seed: solvent ratio, 420 W microwave power, and 0.125 mm particle size using hexane as solvent at 60°C. UAE using ultrasound bath with 53 kHz frequency however gave lower yield of 2.74 mg carvone/g after 30 min. The kinetics of extraction was studied using Peleg's model and was corroborated by plotting predicted and experimental values of carvone yield with good agreement.

12.1   INTRODUCTION

The consumption of natural products is increasing as people are becoming aware of the medicinal, nutritional, and physiological benefits of natural products to the health of human beings. In addition, the use of natural products ensures fewer side effects than when using synthetic products. Natural product extracts find applications in varied fields like pharmaceuticals, food, textiles, and cosmetics. They are also used as drugs, vitamin supplements, preserving agents, chemical standards, edible oils, and fats. Parts of plants such as roots, barks, leaves, and seeds can be used as sources of natural products.[1] The various types of natural products commonly encountered are fatty acids, waxes, terpenoids, steroids, phenolics, alkaloids, polyacetylenes, essential oils, and glycosidic derivatives.[2]

The oldest medical text, dating back to 2600 B.C. in Mesopotamia, contains uses of almost 1000 plants and substances derived from them. Hippocrates of Cos (460–377 B.C.), the Greek physician, derived the use of almost 400 natural substances and portrayed their utilization in his book *Corpus Hippocraticum*.[3] Indian Ayurvedic treatise *Charaka Samhita* (900 B.C.) contains 341 plant-inferred medications. *Sushruta Samhita* (600 B.C.) contains references to 395 medicinal plants and 57 products derived from animal sources.[4]

Carvone exists in two enantiomeric forms, S-(+) carvone and R-(−) carvone; both are naturally occurring compounds. Caraway seed oil contains S-(+) carvone as major component, whereas R-(−) carvone is found in spearmint oil.[5] Carvone is used to inhibit sprouting of potatoes and is used as fungistatic as well as bacteriostatic.[6] It is also used as an insecticide, mosquito repellent, and perfume ingredient, for flavoring of food stuff, and as a starting material for the synthesis of different compounds of high medical relevancy.[7] The main source of carvone recovery is caraway seed, and the recovery has been carried out by a number of researchers using conventional extraction techniques such as steam distillation, hydrodistillation, Soxhlet extraction, etc.[8] Non-conventional techniques that have been used to extract caraway essential oil are supercritical fluid extraction,[9–11] microwave-assisted extraction (MAE),[12] and ultrasound-assisted extraction (UAE).[13]

Steam distillation was used to extract carvone from spearmint (*Mentha spicata* L.)[14] and dill seeds.[15–17] Kokkini et al. and Mkaddem et al. used hydrodistillation for extraction of carvone from spearmint (*Mentha spicata* L.) and *Morelia Viridis*, respectively.[18,19] Dill seed extraction was carried out by Chemat and Esveld and Nautiyal and Tiwari using ultrasound/MAE and supercritical fluid extraction, respectively.[20,21] There are annual and biennial caraway plants; however, the biennial caraway plants are more widely cultivated as they are more productive. The biennial seeds contain 3–7% oil while the annual seeds contain 2–3% oil.[7] Caraway has been cultivated in various countries such as the Netherlands, Denmark, Hungary, Poland, Germany, Czech Republic, Romania, Norway, Ukraine, India, Israel, and Egypt, but it is native to Western Asia, Europe, and Northern Africa.[22] About 30 compounds have been identified from caraway oil, with carvone and limonene making up to 95% of essential oil.[23] Caraway has trace amounts of acetaldehyde, furfural, carveol, pinene, thujone, and camphene, and also contains lipids (13–21%), nitrogen compounds (25–35%), fiber (13–19%), and water (9–13%).[24]

Chemat et al.[12] have used MAE to extract carvone and limonene from caraway seeds. In this work, around 19 mg/g of carvone yield was obtained. However, the yield obtained varies with the variety of plants used, that is, annual and biennial and the region from which it is obtained. Andras et al.[8] used microwave-assisted hydrodistillation for the same purpose.

The conventional techniques of extraction have been used in industry for quite some time. The disadvantages of these processes are that they require longer time, consume more energy, and result in lower yields. Soxhlet extraction is known to take 3 h to almost 24 h for completion and it also requires greater amount of solvents. Hence new techniques such as supercritical fluid extraction, ultrasound and MAE, pressurized liquid extraction, etc., were developed that will increase extraction as well as reduce the solvent consumption.

UAE involves phenomenon of cavitation which is formation of small bubbles. When a bubble implodes near a surface, the cell walls break and this facilitates the diffusion of components from inside of cell to the solvent.[25] One of the notable advantages of ultrasound is that it can be performed at lower temperature. This is particularly useful while extracting thermolabile compounds which may be altered under Soxhlet extraction and which involves the use of high temperature.[26] MAE uses microwaves to heat the raw materials. In microwave heating, the microwaves penetrate the pores and the electromagnetic energy of microwaves is converted to heat by ionic conduction and dipole rotation. Since microwaves penetrate inside the cell, the heating is very rapid and consequently the compounds are also extracted quickly in comparison with the Soxhlet extraction which involves convective type of heating.[27] MAE has attracted attention in field such as medicinal plant research due to its special heating mechanism, moderate capital cost, and good performance under atmospheric conditions.[28]

The aim of this work was to develop the suitable method for carvone extraction and its comparison with conventional Soxhlet extraction method. Well-known Peleg's model was used to predict various extraction kinetic parameters[29] and the effect of various operating parameters such as solvent type, seed to solvent ratio, microwave power, and particle size has been studied.

12.2 EXPERIMENTAL

12.2.1 PLANT MATERIAL AND CHEMICALS

The caraway seeds were purchased from Jalgaon, Maharashtra, India. Different solvents like hexane, ethanol, methanol, chloroform, and xylene were used. All the solvents used were of analytical grade purchased from Sigma Aldrich Chemicals, Mumbai, and carvone standard was purchased from Merck Specialities Private Ltd., Mumbai.

12.2.2 APPARATUS

Soxhlet apparatus of 250 mL capacity with condenser assembly was used for conventional extraction. Ultrasound bath (Equitron) with internal dimensions of $52 \times 30 \times 13$ cm^3 and tank capacity 20 L with a frequency 53 kHz was used for the UAE experiments. Ultrasound bath was equipped with built in heater, digital temperature controller and indicator. Power variation from 208 to 520 W was done using a built in digital controller. Domestic microwave oven (Raga's Electromagnetic Systems) modified in the laboratory with a provision of water condenser and a built in magnetic stirrer was employed for microwave experiments with maximum power rating of 700 W. The microwave was calibrated for 15 s cycle duty. The time duration of microwave irradiation in each cycle is shown in Table 12.1.

TABLE 12.1 Time of Microwave Irradiation Per Cycle.

Microwave power (W)	Time of irradiation (s) (15 s cycle)
140	4
210	5
280	6
350	7
420	8

12.2.3 RAW MATERIAL PREPARATION

The caraway seeds were ground to a powder using a domestic mixer. The powder was sieved using a sieve shaker to get uniform particle sizes. Sieve shaker with sieves of different mesh sizes (16, 72, and 120) were used.

12.2.4 CARVONE QUANTIFICATION

Standard analytical procedure for carvone analysis reported in the literature was used in this study.[11] The pretreated samples were analyzed using Agilent Technologies Gas Chromatograph HP 7890A equipped with auto-sampler. The separation was carried out on an HP-5 column (cross-linked 5% PH ME siloxane, 30 m $\times$ 320 μm $\times$ 0.25 μm) with nitrogen as carrier gas with flow rate of 2 mL/min, injector temperature 220°C, and detector temperature 240°C. The column temperature was set at 60°C for 1 min. Later the temperature was increased to 220°C at a rate of 40°C/min. After 7 min of run time the sequence was stopped and 1 μL of extract was injected in gas chromatography (GC) column. The chromatograms were processed using Ezchrome elite software.

12.2.5   METHODS

12.2.5.1   SOXHLET EXTRACTION

Extraction was performed using 9 g of crushed caraway seeds of 0.125 mm size with hexane as solvent for a seed to solvent ratio of 1:10 in 250 mL round bottom flask. Temperature of heater was set at 69°C which is the boiling point of hexane. Extraction was stopped after 3 h. Samples were collected after regular intervals.

12.2.5.2   ULTRASOUND-ASSISTED EXTRACTION

UAE was done by using ultrasound bath. Three grams crushed caraway seeds of 0.125 mm size were placed with 30 mL hexane in 250 mL conical flask. Power of ultrasonic bath was 208 W. Extraction was carried out for 30 min and the samples were withdrawn at regular intervals for carvone quantification.

12.2.5.3   MICROWAVE-ASSISTED EXTRACTION

MAE was carried out by placing 3 g of crushed caraway seeds in a 250 mL conical flask along with 30 mL of hexane as solvent. The conical flask was connected to a condenser at the top. Water was used with condenser to condense the vapors back in the flask. Effect of microwave power on extraction of carvone was studied by varying microwave power from 140 to 420 W at 70 W interval. The effect of solid-to-solvent ratio on carvone yield was also studied by changing it from 1:10 to 1:30 keeping all other parameters unchanged. To investigate the effect of particle size on carvone yield, particles of 1.19, 0.210, and 0.125 mm size were used. Samples were collected at regular intervals.

12.3   RESULTS AND DISCUSSION

12.3.1   KINETIC MODEL

Peleg's kinetic model, which is a semi empirical model (eq 12.1), is used to describe the extraction curves of biological materials from plant sources as they show similarity to the sorption curves.[9]

$$C_t = C_o + \frac{t}{K_1 + K_2.t} \tag{12.1}$$

where C_t is the concentration of target solute at time t (mg/g powder), K_1 is Peleg's rate constant (min·g/mg), and K_2 is Peleg's capacity constant (g/mg). The term C_o can be omitted from eq 12.1 as it expresses the initial concentration of target solute in the extraction solvent which is zero. Concentration of solute can be measured at corresponding time, and its behavior is explained using the modified Peleg's equation (eq 12.2):

$$C_t = \frac{t}{K_1 + K_2.t} \tag{12.2}$$

Peleg's rate constant (K_1) and Peleg's capacity constant (K_2) can be obtained from the graph of $1/C_t$ vs. $1/t$.

12.3.2 COMPARISON OF SOXHLET, ULTRASOUND, AND MICROWAVE-ASSISTED EXTRACTION

To select the most suitable method for extracting carvone from caraway seeds, initially the extraction was carried out using Soxhlet apparatus, ultrasound bath, and microwave.

12.3.2.1 SOXHLET EXTRACTION

Soxhlet extraction was carried out using 9 g of crushed caraway seeds and 90 mL hexane at 69°C temperature and the results are reported in Figure 12.1. Extraction was carried out for 3 h which resulted in 10.03 mg carvone/g of seeds. This was considered as the 100% yield for the comparison purpose with the other methods. Even after increasing the extraction time to 5 h (results not reported here), no increase in extraction yield was observed. The yield obtained here is less than what was reported by Chemat et al. which was 17.12 mg carvone/g.[12] This is because for countries like India, where climate is hot and dry, inferior quality of caraway seeds are produced which suggests that sufficient moisture availability is necessary for higher yield.[30]

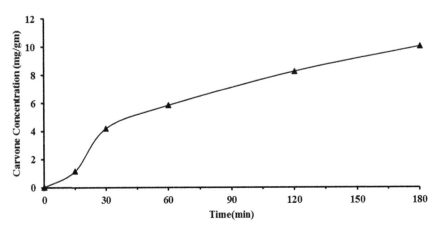

FIGURE 12.1　Carvone concentration vs. time for Soxhlet extraction (temperature—69°C, solvent—hexane, solid/solvent ratio—1:10, and particle size—0.125 mm).

12.3.2.2　EXTRACTION USING ULTRASONIC BATH

UAE was carried out using ultrasound bath operating at 208 W and 53 kHz frequency. For this, 3 g seeds of 0.125 mm size were added with 30 mL hexane in 250 mL conical flask. No cooling provision to carry away heat generated during sonication was used. Extraction was stopped after 30 min and only one sample at the end of 30 min was taken. The temperature range of the mixture was 32–45°C for non-controlled temperature in ultrasound experiment (without cooling). After 30 min, ultrasound irradiations yield of 2.74 mg carvone/g was obtained.

12.3.2.3　MICROWAVE-ASSISTED EXTRACTION

MAE was carried out using 3 g of seeds and 30 mL solvent in 250 mL round bottom flask at 210 W microwave power. After 30 min of microwave irradiation, a yield of 5.66 mg carvone/g was obtained. The values of extraction yield of carvone for three methods for 30 min experimental time are listed in Table 12.2 and shown in Figure 12.2.

　　Comparing the results of these three methods for 30 min of experimental time, it can be seen that (Table 12.2) yield obtained with MAE is almost 1.3 times higher than that obtained with Soxhlet extraction and 2.1 times more than that obtained with ultrasound bath.

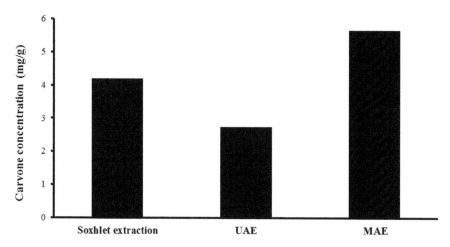

FIGURE 12.2 Comparison of yield obtained with Soxhlet extraction, UAE, and MAE after 30 min of extraction (solvent—hexane, seed to solvent ratio—1:10, and particle size—0.125 mm).

TABLE 12.2 Comparison of Yield Obtained by Soxhlet, Microwave, and Ultrasound-Assisted Extraction (solvent—hexane, solid/solvent ratio = 1:10, and time duration = 30 min).

Extraction technique	Yield of carvone (mg/g)	Extraction conditions
Soxhlet	4.20	69°C
Ultrasound	2.74	208 W, 32–45°C
Microwave	5.66	210 W, 30–60°C

From these results, it can be seen that MAE resulted in highest amount of carvone extraction followed by Soxhlet extraction and UAE. Soxhlet extraction was carried out at higher temperature (69°C) whereas MAE was carried out at 60°C. There will be definite amount of energy savings with MAE as compared to Soxhlet extraction. At the same time ultrasound method resulted in lowest carvone extraction (2.74 mg/g) which may be because of ultrasound bath which induces indirect sonication. As a result, lower extraction yield is observed. Since the amount of carvone extracted is highest with MAE than the Soxhlet extraction and ultrasound method, MAE was found to be the best suited for carvone extraction. Hence the parametric study was carried out using MAE.

The results obtained with MAE with respect to various parameters like microwave power, time of irradiation, solvents, particle size, and solid/solvent ratio are presented in the following section.

12.3.3 EFFECT OF SOLVENT TYPE

Selecting an appropriate solvent is very important to develop an optimized extraction process. To select the best solvent for the extraction, the experiments were performed with hexane, methanol, ethanol, chloroform, and xylene. Other extraction conditions were 210 W microwave power, 1:10 solid-to-solvent ratio, and 0.125 mm particle size. The results are reported in Figure 12.3.

The results show that hexane gave the highest yield of 5.66 mg carvone/g. The second-best solvent was methanol with a yield of 5.41 mg carvone/g. Methanol has higher value of dielectric constant than hexane, so it gets rapidly heated as compared to hexane which has very low value of dielectric constant. Due to this, there is a risk of explosion using methanol in laboratory scale microwave oven; hence it was not used for further experiments. Use of chloroform, even though it gives a somewhat similar yield (5.05 mg/g), creates problems during filtration of extract because the plant material also comes with solvent due to higher density of chloroform. This can result in increased cost of post extraction treatment and increased time for filtration. A yield of 4.69 mg carvone/g was obtained with xylene. As the boiling point of xylene is high it requires more energy during solvent recovery.

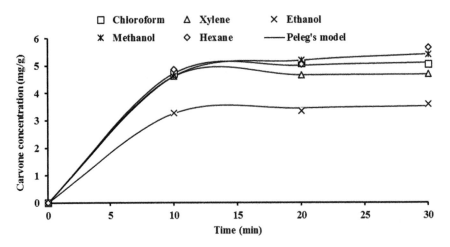

FIGURE 12.3 Effect of different solvents on the kinetics of MAE of carvone from caraway seeds and Peleg's model fitting (solid/solvent ratio—1:10, particle size—0.125 mm, and power—210 W).

The least yield of 3.59 mg carvone/g was achieved with the use of ethanol which is probably due to low solubility of carvone in ethanol. Based on the above results, hexane was selected as the best solvent for the present case. Hexane is available at cheaper rate and the recovery of hexane from the extract is easier using rotary vacuum distillation. As a result, it consumes less energy and makes it desirable to use hexane at industrial scale where costs associated with energy consumption are major contributors to overall economic viability and competitiveness of extraction process. Hence all the experiments were performed using hexane.

The values of concentration of carvone for different solvents corresponding to different times were predicted by Peleg's kinetic model. Values for Peleg's kinetic model are presented in Table 12.3 and plotted in Figure 12.3, respectively, indicating acceptable concurrence with experimental values.

TABLE 12.3 Effect of Different Solvents on the Kinetic Parameters, i.e., Peleg's Rate Constant, K_1 (min·g/g) and Peleg's Capacity Constant, K_2 (g/mg); and Comparison of Experimental and Calculated C_{eq} (mg carvone/g).

Solvent	Experimental C_{eq} (mg carvone/g)	K_1 (min·g/ mg)	K_2 (g/mg)	Calculated C_{eq} (mg carvone/g)	RMSD
Methanol	5.41	0.48	0.17	5.41	0.006
Xylene	4.69	0.05	0.21	4.68	0.008
Chloroform	5.05	0.24	0.19	5.11	0.060
Hexane	5.66	0.38	0.17	5.46	0.179
Ethanol	3.59	0.34	0.27	3.52	0.071

12.3.4 EFFECT OF SOLID-TO-SOLVENT RATIO

Five different solid-to-solvent ratios starting from 1:10 to 1:30 were used to study the effect of solid-to-solvent ratio on extraction yield of carvone. The other experimental conditions were microwave power of 210 W and 30 min of irradiation. The results are shown in Figure 12.4. From the figure, it can be seen that there is no appreciable difference in the yield of carvone with changing solid-to-solvent ratios. The yield of 5.66 mg carvone/g was obtained with 1:10 seed to solvent ratio which increased marginally with the increasing quantity of solvent till the ratio of 1:25 and after that the yield dropped to 5.28 mg carvone/g for the ratio of 1:30.

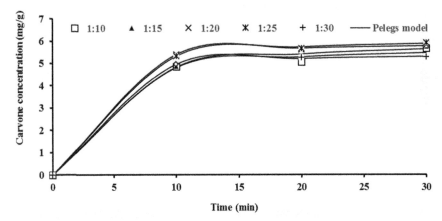

FIGURE 12.4 Effect of different seed to solvent ratios on the kinetics of microwave-assisted extraction of carvone from caraway seeds and Peleg's model fitting (solvent ratio—hexane, particle size—0.125 mm, and power—210 W).

The probable reason for the initial increase in carvone yield may be because of the presence of large concentration gradient at higher solid-to-solvent ratio which facilitates the mass transfer. However, this effect is more substantial at smaller solid-to-solvent ratios. Also, larger quantity of solvent does not change the concentration gradient as the restriction to the mass transfer is mainly associated with the solid interior.[31]

Similar results are reported by Vetal et al.[31] in their work on extraction of ursolic acid from *Ocimum sanctum* by using ultrasound. In this work, increase in yield of ursolic acid with increase in solid-to-solvent ratio from 1:10 to 1:30 was observed and after that yield decreased as the ratio increased from 1:30 to 1:50. In another work of Vetal et al.[32] for MAE of ursolic acid and oleanolic acid from *O. sanctum* the extraction yield of ursolic acid and oleanolic acid increased with the increase in ratio from 1:10 to 1:30 and further it decreased. The decrease in extraction yield with increased amount of solvent may be because large amount of solvent absorbs more microwave energy; therefore, sufficient microwave energy may not be available for the effective disruption of the plant cells. Hence a solid-to-solvent ratio of 1:10 was chosen as optimum as this will require less solvent which in turn will reduce the solvent losses during solvent recovery. The predicted model values and experimental values are given in Table 12.4 and plotted in Figure 12.4.

TABLE 12.4 Effect of Different Solid:Solvent Ratio on the Kinetic Parameters, i.e., Peleg's Rate Constant, K_1 (min·g/g) and Peleg's Capacity Constant, K_2 (g/mg) and Comparison of Experimental and Calculated C_{eq} (mg carvone/g).

Solid: solvent ratio	Experimental C_{eq} (mg carvone/g)	K_1 (min·g/mg)	K_2 (g/mg)	Calculated C_{eq} (mg carvone/g)	RMSD
1:10	5.66	0.38	0.17	5.46	0.179
1:15	5.75	0.44	0.16	5.64	0.099
1:20	5.81	0.18	0.17	5.77	0.041
1:25	5.91	0.27	0.16	5.87	0.039
1:30	5.28	0.20	0.18	5.31	0.025

12.3.5 EFFECT OF MICROWAVE POWER

The effect of microwave power was studied by using five different values of power—140, 210, 280, 350, and 420 W. For these experiments 3 g of caraway seeds and 30 mL hexane with 0.125 mm particle size was used. Figure 12.5 represents the effect of microwave power   on extraction of carvone with respect to time. With increase in microwave power from 140 to 350 W, moderate increase in yield (5.32—6.26 mg carvone/g) was observed. Initially the extraction increased by small amounts as power of microwaves was increased. The higher extraction yield can be ascribed to ionic conduction and dipole rotation which generates molecular movement and heat. Higher extraction yield at higher power is possibly due to transfer of more electromagnetic energy to the extraction medium in less time which causes overall enhancement in the extraction efficiency.[32] Chen and Spiro[33] reported that the extraction rate increased with the increase in microwave power up to a certain level during extraction of rosemary leaves using MAE. Chemat et al.[12] in their study on MAE of terpenes from caraway seeds demonstrated that extraction yield increased with power from 50 to 150 W and after that no increase in yield was noted. However, during our experiments, which involved the same raw material and extraction method, carvone yield was found to increase for the range of 140–420 W. This difference can be due to the use of different variety of caraway used in our experiment.

The highest yield of 8.77 mg carvone/g was obtained at 420 W power which is approximately equal to 87% as compared to 3 h Soxhlet extraction (100%). Further increase in power after 420 W resulted in considerably low yield which can be attributed to loss of material due to degradation of compound (results not reported here). Thus, we can deduce that the difference in results in various studies is due to the different nature of matrix used and which component is being extracted. Though the highest yield of 8.77 mg/g was obtained at 420 W after 30 min of irradiation, still power of 210 W was used as optimum value for further experiments. The reason for this selection can be explained on the basis that with double increment in power input from 210 to 420 W the proportionate increase in the yield was not achieved. Peleg's kinetic model values are presented in Table 12.5 and plotted in Figure 12.5, respectively, showing satisfactory agreement with experimental values.

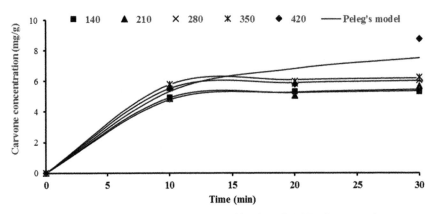

FIGURE 12.5   Effect of microwave power on kinetics of MAE of carvone from caraway seeds and Peleg's model fitting (solvent—hexane, solid/solvent ratio—1:10, and particle size—0.125 mm).

TABLE 12.5   Effect of Different Microwave Power on Kinetic Parameters, i.e., Peleg's Rate Constant, K_1 (min·g/mg) and Peleg's Capacity Constant, K_2 (g/mg) and comparison of Conventional and Calculated C_{eq} (mg/g).

Power (W)	Experimental C_{eq} (mg carvone/g)	K_1 (min·g/mg)	K_2 (g/mg)	Calculated C_{eq} (mg carvone/g)	RMSD
140	5.32	0.24	0.18	5.36	0.043
210	5.66	0.38	0.17	5.46	0.179
280	6.07	0.23	0.16	6.03	0.042
350	6.26	0.17	0.16	6.20	0.064
420	8.77	0.82	0.11	7.51	0.942

12.3.6   EFFECT OF PARTICLE SIZE

To study the effect of raw material particle size three different particle sizes, namely 1.19, 0.210, and 0.125 mm were used. For this purpose, 3 g of seeds and 30 mL hexane solvent were irradiated with microwaves at 210 W for 30 min. Samples were taken after 10 min intervals and the results are depicted in Figure 12.6. With decrease in the particle size from 1.19 to 0.125 mm, extraction yield increased considerably from 1.4 to 5.66 mg carvone/g, respectively.

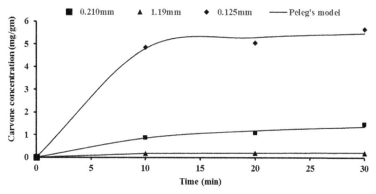

FIGURE 12.6   Effect of different particle sizes on the kinetics of MAE of carvone from caraway seeds and Peleg's model fitting (solvent—hexane, solid/solvent ratio—1:10, and power—210 W).

From Figure 12.6, it can be observed that the highest yield was obtained for the smallest particle size of 0.125 mm which may be due to the large surface area available to provide contact between sample and solvent, facilitating easy release of components in the solvent. It is reported that extraction yield increases as the particles size is reduced.[34] For large size particles more time is needed for diffusion of essential components from cells into the solvent. Gião et al.[35] reported higher yield with decrease in size of particles while extracting antioxidant from *Satureja montana*. Reducing the particle size results in increase of the superficial area which in turn increases the extraction rate. It also decreases intra particle diffusion path which results in more efficient extraction. However, Coelho et al.[36] reported no significant differences in yield (81–85%) for change in particle size from 0.55 to 0.35 mm.[36] With coarser particles of 0.21 and 1.19 mm, comparatively less extraction yield of carvone was obtained, as seen in Figure 12.6. Based on the above results 0.125 mm particle size was chosen for further experiments. The predicted model values and experimental values are presented in Table 12.6

and plotted in Figure 12.6, which show satisfactory agreement with experimental values.

TABLE 12.6 Effect of Different Particle Sizes on the Kinetic Parameters, i.e., Peleg's Rate Constant, K_1 (min·g/g) and Peleg's Capacity Constant, K_2 (g/mg) and Comparison of Experimental and Calculated C_{eq} (mg carvone/g).

Particle size	Experimental C_{eq} (mg carvone/g)	K_1 (min·g/mg)	K_2 (g/mg)	Calculated C_{eq} (mg carvone/g)	RMSD
0.125	5.66	0.38	0.17	5.463	0.179
0.210	1.46	6.57	0.52	1.36	0.082
1.19	0.20	10.84	4.49	0.21	0.003

12.4 CONCLUSION

The extraction of carvone using ultrasound and microwave was studied and compared with conventional method of Soxhlet extraction. Considering the yield obtained using Soxhlet extraction for 3 h to be 100%, around 87% extraction yield is obtained with MAE in 30 min which is one-sixth of the time required for conventional Soxhlet extraction. UAE on the other hand resulted in nearly 27% yield in 30 min. From the results obtained it can be concluded that MAE is a better technique than UAE and conventional Soxhlet extraction for carvone extraction from caraway seeds. The optimum conditions for the maximum yield of 8.77 mg carvone/g using hexane as solvent were 420 W power input, seed-to-solvent ratio of 1:10, and particle size of 0.125 mm 60°C. The kinetics of extraction was studied using Peleg's model and was corroborated by plotting predicted and experimental values of carvone yield with good agreement.

KEYWORDS

- carvone
- caraway seed
- microwave-assisted extraction
- ultrasound-assisted extraction
- Soxhlet extraction

REFERENCES

1. Cavalcanti, R. N.; Forster-Carneiro, T.; Gomes, M. T. M. S.; Rostagno, M. A.; Prado, J. M.; Meireles, M. A. A. Uses and Applications of Extracts from Natural Sources. *Natural Product Extraction Principles and Applications;* Rostagno, M. A., Prado, J. M., Eds.; RCS Publishing: Cambridge, 2013; p 1.
2. Seidel, V. Initial and Bulk Extraction. *Natural Products Isolation;* 2nd ed.; Sarker, S. D., Latif, Z., Gray, A. I., Eds.; Humana Press: Totowa, NJ, 2005; p 29.
3. Ji, H. F.; Li, X. J.; Zhang, H. Y. Natural Products and Drug Discovery. *EMBO Rep.* **2009,** *10,* 194–200.
4. Dev, S. Ancient-Modern Concordance in Ayurvedic Plants: Some Examples. *Environ. Health Persp.* **1999,** *107,* 783–789.
5. Lambert, J.; Compton, R. N.; Crawford, T. D. The Optical Activity of Carvone: A Theoretical and Experimental Investigation. *J. Chem. Phys.* **2012,** *136,* 114512.
6. Oosterhaven, K.; Poolman, B.; Smid, E. J. S-Carvone as a Natural Potato Sprout Inhibiting, Fungistatic and Bacteristatic Compound. *Ind. Crop. Prod.* **1995,** *4,* 23–31.
7. Decarvalho, C.; Dafonseca, M. Carvone: Why and how should One Bother to Produce this Terpene. *Food Chem.* **2006,** *95,* 413–422.
8. Andras, C. D.; Salamon, R. V.; Barabas, I.; Volf, I.; Szep, A. Influence of Extraction Methods on Caraway (*Carum carvi* l.) Essential Oil Yield and Carvone/Limonene Ratio. *Environ. Eng. Manag. J.* **2015,** *14,* 341–347.
9. Baysal, T.; Starmans, D. A. J. Supercritical Carbon Dioxide Extraction of Carvone and Limonene from Caraway Seed. *J. Supercrit. Fluid.* **1999,** *14,* 225–234.
10. Sovova, H.; Komers, R.; Kucera, J.; Jez, J. Supercritical Carbon-Dioxide Extraction of Caraway Essential Oil. *Chem. Eng. Sci.* **1994,** *49,* 2499–2505.
11. Sedláková, J.; Kocourková, B.; Lojková, L.; Kubáň, V. Determination of Essential Oil Content in Caraway (*Carum carvi* L.) Species by Means of Supercritical Fluid Extraction. *Plant. Soil Environ.* **2003,** *49,* 277–282.
12. Chemat, S.; Ait-Amar, H.; Lagha, A.; Esveld, D. C. Microwave-Assisted Extraction Kinetics of Terpenes from Caraway Seeds. *Chem. Eng. Process.* **2005,** *44,* 1320–1326.
13. Chemat, S.; Lagha, A.; Ait-Amar, H.; Bartels, P. V.; Chemat, F. Comparison of Conventional and Ultrasound-Assisted Extraction of Carvone and Limonene from Caraway Seeds. *Flavour Frag. J.* **2004,** *19,* 188–195.
14. Younis, Y.; Beshir, S. M. Carvone-Rich Essential Oils from *Mentha longifolia* (L.) Huds. ssp. schimperi *Briq.* and *Mentha spicata* L. Grown in Sudan. *J. Essent. Oil Res.* **2004,** *16,* 539–541.
15. Embong, M. B.; Hadziyev, D.; Molnar, S. Essential Oils from Spices Grown in Alberta Dill Seed Oil, *Anethum graveolens,* L. (Umbelliferae). *Can. Inst. Food Sci. Tech. J.* **1977,** *10,* 208–214.
16. Bailer, J.; Aichinger, T.; Hackl, G.; de Hueber, K.; Dachler, M. Essential Oil Content and Composition in Commercially Available Dill Cultivars in Comparison to Caraway. *Ind. Crop. Prod.* **2001,** *14,* 229–239.
17. Callan, N. W.; Johnson, D. L.; Westcott, M. P.; Welty, L. E. Herb and Oil Composition of Dill (*Anethum graveolens* L.): Effects of Crop Maturity and Plant Density. *Ind. Crop. Prod.* **2006,** *25,* 282–287.
18. Kokkini, S.; Karousou, R.; Lanaras, T. Essential Oils of Spearmint (Carvone-rich) Plants from the Island of Crete (Greece). *Biochem. Syst. Ecol.* **1995,** *23,* 425–430.

19. Mkaddem, M.; Bouajila, J.; Ennajar, M.; Lebrihi, A.; Mathieu, F.; Romdhane, M. Chemical Composition and Antimicrobial and Antioxidant Activities of Mentha (*Longifolia* L. and *Viridis*) Essential Oils. *J. Food Sci. Tech. Mys.* **2009,** *74,* 358–363.

20. Chemat, S.; Esveld, E. D. C. Contribution of Microwaves or Ultrasonics on Carvone and Limonene Recovery from Dill Fruits (*Anethum graveolens* L.). *Innov. Food Sci. Emerg. Technol.* **2013,** *17,* 114–119.

21. Nautiyal, O. P.; Tiwari, K. K. Extraction of Dill Seed Oil (*Anethum sowa*) Using Supercritical Carbon Dioxide and Comparison with Hydrodistillation. *Ind. Eng. Chem. Res.* **2011,** *50,* 5723–5726.

22. Bouwmeester, H. J.; Gershenzon, J.; Konings, M. C.; Croteau, R. Biosynthesis of the Monoterpenes Limonene and Carvone in the Fruit of Caraway. I. Demonstration of Enzyme Activities and their Changes with Development. *Plant Physiol.* **1998,** *117,* 901–912.

23. Kallio, H.; Kerrola, K.; Alhonmaki, P. Carvone and Limonene in Caraway Fruits (*Carum carvi* L) Analyzed by Supercritical Carbon-Dioxide Extraction Gas-Chromatography. *J. Agric. Food Chem.* **1994,** *42,* 2478–2485.

24. Sedlakova, J.; Kocourkova, B.; Kuban, V. Determination of Essential Oil Content and Composition in Caraway Seeds. *Czech. J. Food Sci.* **2001,** *19,* 31–36.

25. Pico, Y. Ultrasound-Assisted Extraction for Food and Environmental Samples. *Trends Food Sci. Technol.* **2014,** *43,* 84–99.

26. Romdhane, M.; Gourdon, C. Investigation in Solid-Liquid Extraction: Influence of Ultrasound. *Chem. Eng. J.* **2002,** *87,* 11–19.

27. Kaufmann, B.; Christen, P. Recent Extraction Techniques for Natural Products: Microwave-Assisted Extraction and Pressurised Solvent Extraction. *Phytochem. Anal.* **2002,** *13,* 105–113.

28. Chan, C. H.; Yusoff, R.; Ngoh, G. C.; Kung, F.W. Microwave-Assisted Extractions of Active Ingredients from Plants. *J. Chromatogr. A.* **2011,** *1218,* 6213–6225.

29. Karacabey, E.; Bayindirli, L.; Artik, N.; Mazza, G. Modeling Solid-Liquid Extraction Kinetics of Trans-Resveratrol and Trans-E-Viniferin from Grape Cane. *J. Food Process Eng.* **2011,** *36,* 1–10.

30. Toxopeus, H.; Bouwmeester, H. J. Improvement of Caraway Essential Oil and Carvone Production in the Netherlands. *Ind. Crop Prod.* **1993,** *1,* 295–301.

31. Vetal, M. D.; Lade, V. G.; Rathod, V. K. Extraction of Ursolic Acid from *Ocimum sanctum* by Ultrasound: Process Intensification and Kinetic Studies. *Chem. Eng. Process.* **2013,** *69,* 24–30.

32. Vetal, M. D.; Chavan, R. S.; Rathod, V. K. Microwave Assisted Extraction of Ursolic Acid and Oleanolic Acid from *Ocimum sanctum*. *Biotechnol. Bioprocess Eng.* **2014,** *19,* 720–726.

33. Chen, S. S.; Spiro, M. Kinetics of Microwave Extraction of Rosemary Leaves in Hexane, Ethanol and a Hexane + Ethanol Mixture. *Flavour Frag. J.* **1995,** *10,* 101–112.

34. Kaufmann, B.; Christen, P.; Veuthey, J. L. Parameters Affecting Microwave-Assisted Extraction of with Anolides. *Phytochem. Analysis.* **2001,** *12,* 327–331.

35. Gião, M. S.; Pereira, C. I.; Fonseca, S. C.; Pintado, M. E.; Malcata, F. X. Effect of Particle Size upon the Extent of Extraction of Antioxidant Power from the Plants *Agrimonia eupatoria, Salviasp.* and *Satureja montana*. *Food Chem.* **2009,** *117,* 412–416.

36. Coelho, J. A. P.; Pereira, A. P.; Mendes, R. L.; Palavra, A. M. F. Supercritical Carbon Dioxide Extraction of *Foeniculum vulgare* volatile Oil. *Flavour Frag. J.* **2003,** *18,* 316–319.

CHAPTER 13

KINETIC MODEL FOR EXTRACTION OF BETULINIC ACID BY BATCH EXTRACTION FROM LEAVES OF *SYZYGIUM CUMINI* (JAMUN)

S. V. ADMANE[1,*], S. G. GAIKWAD[2], and S. M. CHAVAN[3]

[1]*Department of Chemical Engineering, MIT Academy of Engineering, Alandi, Pune, Maharashtra, India*

[2]*CEPD Department, National Chemical Laboratory, Pune, Maharashtra, India*

[3]*Department of Chemical Engineering, Sinhgad College of Engineering, Vadgaon, Pune, Maharashtra, India*

Corresponding author. E-mail: shraddhaadmane@gmail.com

CONTENTS

ABSTRACT

Plants are one of the most important sources of medicines. *Syzygium cumini* known as "jamun" in Hindi which is found in throughout India. The leaves are having promising therapeutic value with its various phyto constituents such as tannin, alkaloids, flavonoids, terpenoids, fatty acid, phenol, minerals, carbohydrates, and vitamins. All these phyto constituents are analyzed by qualitative analysis method. Betulinic acid is a naturally occurring penta-cyclic triterpenoid compound. The medicinal application of betulinic acid is available for diabetes, cancer cell, and HIV. Betulinic acid shows more solubility for methanol. For extraction of betulinic acid, different extraction techniques like batch extraction and Soxhlet extraction techniques are used. The effect of various parameters like particle size, agitations speed, temperature, and solid loading is studied. To explain the extraction of biological material from plant sources different model equations are used which describes sorption isotherms of material. The experimental data of solid loading and temperature are best fitted by using second order Peleg's model. From this model the linearized Arrhenius equation was obtained and temperature dependency was 0.567 kJ/mol. The aim of this study is to find out the best kinetic model for extraction and experimental data shows the good agreement of betulinic acid with methanol which justifies batch extraction process.

13.1 INTRODUCTION

Plant is a source of medicines. The medicinal plants have high therapeutic importance. In traditional system of medicines, the Indian medicinal plants have been used for the treatment of different diseases. *Syzygium cumini* (family—Myrtaceae) has several medicinal applications. *S. cumini*, known as "Jamun" in Hindi, is found throughout India. The leaves of jamun plants contain many active ingredients which have many medicinal applications. The jamun leaf contains carbohydrate 14%, protein 0.15–0.30 g, fiber 0.30–0.90%, calcium 8.30–15.00 mg, potassium 55.00 mg, phosphorus 15–16.20 mg, magnesium 35.00 mg, iron 1.20–1.60 mg, folic acid 3.00 µg.[1,2] The leaves contain sitosterol, betulinic acid, and crategolic (maslinic) acid.

Pentacyclic triterpenoid compound, known as betulinic acid, possesses various medicinal as well as biological properties such as anti-HIV activity and hypoglycemic, anti-bacterial, and anti-diarrheal effects.[3] The growth of Gram positive and Gram negative bacteria is inhibited by methanol and water extract of *S. cumini*. The growth and colony-forming ability of all human melanoma cell lines are suppressed by betulinic acid. The derivatives of betulinic acid such as betulin, lupeol 3-oxo-23-hydroxybetulinic acid, and 23-hydroxybetulinic acid are also used as medicinal applications.[4] The percent of betulinic acid is present approximately about 2.1% in the leaves of *S. cumini*. Betulinic acid extracted from the leaves of *Vitex negundo* and Tanzanian tree shows anti-bacterial activity against *Bacillus subtilis* and anti-malarial activity against *Plasmodium falciparum,* respectively. Recently betulinic acid is used for the medicines of diabetes, cancer cells, and HIV. Betulinic acid is generally water soluble but it also shows solubility with some alcohols such as methanol and ethanol. But it is more soluble in methanol than ethanol.

| (a) | (b) | (c) | (d) | (e) |

FIGURE 13.1   Qualitative analysis tests: (a) tannin, (b) phytosterol, (c) protein, (d) steroid, and (e) vitamin C.

13.2   MATERIALS AND METHODS

13.2.1   MATERIALS

S. cumini (Jamun) leaves were collected from nearby area of Dhayri, Pune district, Maharashtra, India. They were washed with water and air-dried. The washed leaves were crushed and grinded. The powdered material was segregated by lab screening method and 210 micron size is selected for further experimentation. The qualitative analysis tests for estimation of tannin, phytosterol, protein, steroid, and vitamin C have been done and results are shown in Figure 13.1.

13.2.2 BATCH EXTRACTION SETUP

The extraction of betulinic acid is carried out with batch extraction setup (Fig. 13.2). It consists of a 250 mL reactor and a four-blade agitator connected with motor (REMI maximum speed of 3000 rpm). Methanol is used as the solvent for extraction. The powder form of leaves having a particle size 210 micron is used for extraction setup.

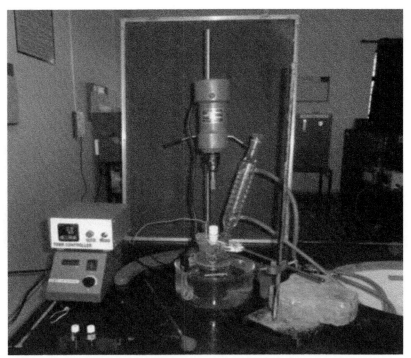

FIGURE 13.2 Experimental setup for batch extraction process of betulinic acid.

13.2.3 ANALYSIS

The extracted samples are analyzed on UV spectrophotometer. Methanol is used as a reference solution for UV spectrophotometer (Thermo Fisher).

13.2.4 MATHEMATICAL MODEL

Many researchers have used various models to describe solid–liquid extraction process. Peleg introduced semi empirical kinetic model to explain extraction curves of biological materials from plant sources. The following model equation was used to describe the sorption isotherms of material:

$$C_t = C_0 + \frac{t}{K_1 + K_2 t} \tag{13.1}$$

where C_t is the concentration of betulinic acid at time t (mg of BA/g of powder); K_1 is Peleg's rate constant (min g/mg); and K_2 is Peleg's capacity constant (g/mg).

Since fresh solvent was used at the beginning of the extraction, initial concentration of target solution was considered zero. So term C_0 can be excluded from Peleg's equation in our studies. Even though the extraction process shows first order kinetic behavior at the beginning, it later decreases to second order kinetics. Therefore, eq 13.1 can be reduced to eq 13.2 which described the relation between target solute concentrations in extraction solvent versus time.

$$C_t = \frac{t}{K_1 + K_2 t} \tag{13.2}$$

The plot of $1/C_t$ versus $1/t$ gives value of slope as K_1 (Peleg's rate constant) and intercept as K_2 (Peleg's capacity constant).

13.3 RESULTS AND DISCUSSIONS

13.3.1 BATCH EXTRACTION

Fresh leaves of S. cumini were washed with tap water and then kept 46 h for drying. After drying, leaves were cut into small pieces. Two gram of weighted pieces fed to the stirred type batch reactor with 200 mL methanol. The experiment was performed for various adsorbent doses and temperature at constant higher speed (900 rpm).

The extracted samples at known intervals of time were analyzed with UV spectrophotometer for betulinic acid content. From Figure 13.3, it is clear that as adsorbent dose increases, the extraction of betulinic acid also decreases. Similarly as temperature increases (Fig. 13.4), the extraction of betulinic acid also increases.

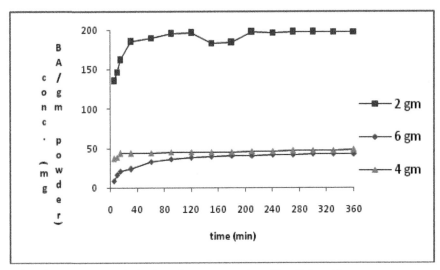

FIGURE 13.3   Effect of adsorbent dose on extraction of betulinic acid.

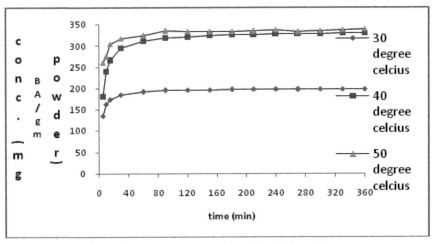

FIGURE 13.4   Effect of temperature on extraction of betulinic acid.

13.3.2 KINETIC MODEL

Second order kinetic model is used for extraction of betulinic acid from *S. cumini*. Figures 13.5–13.10 show the comparison between experimental data with model values of concentration of betulinic acid at various adsorbent doses and at various temperatures, respectively.

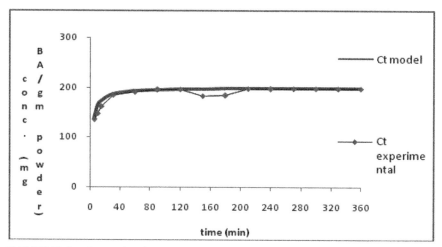

FIGURE 13.5 Comparison between experimental data with model values of betulinic acid concentration at 2 g.

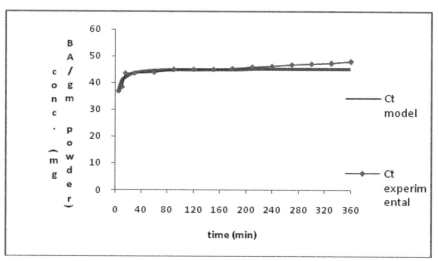

FIGURE 13.6 Comparison between experimental data with model values of betulinic acid concentration at 4 g.

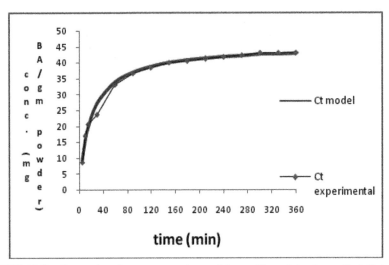

FIGURE 13.7 Comparison between experimental data with model values of betulinic acid concentration at 6 g.

Results above show that when solid loading increases, percent extraction of betulinic acid decreases. But when temperature increases percent extraction of betulinic acid increases. With the increase in the temperature, a higher extraction rate was observed due to direct penetration of bulk solution through leaf matrix.

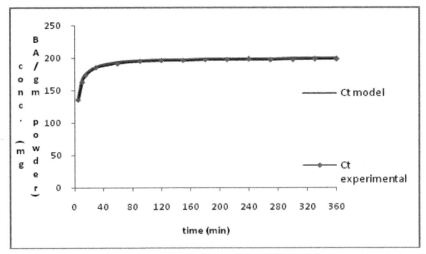

FIGURE 13.8 Comparison between experimental data with model values of betulinic acid concentration at the temperature of 30°C.

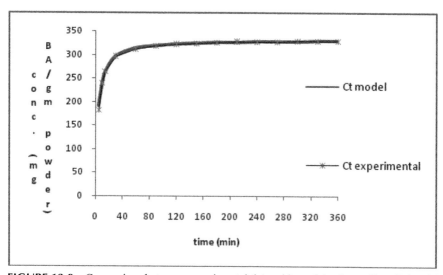

FIGURE 13.9   Comparison between experimental data with model values of betulinic acid concentration at the temperature of 40°C.

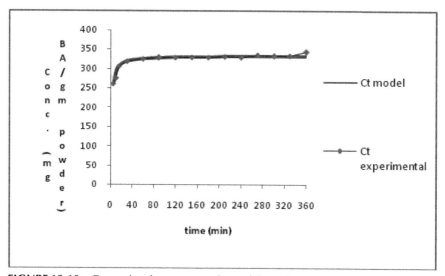

FIGURE 13.10   Comparison between experimental data with model values of betulinic acid concentration at the temperature of 50°C.

13.3.3 MODELING THE EFFECT OF TEMPERATURE AND CONCENTRATION ON THE K1 (PELEG'S RATE CONSTANT)

An analogy between Peleg's model equation with linearized Arrhenius equation described the relation between $1/K_1$ and extraction temperature.

$$\ln\left(\frac{1}{K_1}\right) = \ln(A) - \frac{E_a}{RT} \qquad (13.3)$$

where K_1 is the Peleg's rate constant (min g/mg BA); A is the constant as frequency factor (min^{-1}); E_a is the activation energy (J/mol); R is the universal gas constant (8.314 J/mol k); T is the absolute temperature in K.

The graphical representation between $\ln(1/k_1)$ versus $1/T$ results in a straight line with slope as E_a/R and intercept as $ln(A)$. The experimental activation energy of the batch extraction process was 0.567 kJ/mol.

13.4 CONCLUSION

The extraction of betulinic acid was done by using methanol as a solvent by UV spectrophotometer. The experimentation showed that at low solid loading and at high temperature the rate of extraction is higher. The Peleg's model validation was done by plotting graph between temperature and solid loading with respect to time. For the amount of betulinic acid in extract, the used kinetic model showed good agreement between experimental data and model values.

KEYWORDS

- medicinal plants
- melanoma cell
- qualitative analysis
- extraction rate
- leaf matrix

REFERENCES

1. Swami, S. B.; Thakor, N. J. S.; Patil, M. M.; Haldankar, P. M. Jamun (*Syzygium cumini* (L)): A Review of its Food and Medicinal Uses. *Food Nutr. Sci.* **2012,** *3,* 1100.
2. Fabricant, D. S.; Farnsworth, N. R. The Value of Plants Used in Traditional Medicine for Drug Discovery. *Environ. Health Perspect.* **2001,** *109,* 69.
3. Yogeswari, P.; Sriram, D. Betulinic Acid and its Derivatives: A Review on their Biological Properties. *Curr. Med. Chem.* **2005,** *12,* 657–666.
4. Moghaddam, M. G.; Ahmad F. H.; Kermani, A. S. Biological Activity of Betulinic Acid: A Review. **2012,** *3* (2), 119–123.

CHAPTER 14

REACTIVE EXTRACTION OF PROPIONIC ACID

A. THORGULE* and R. P. UGWEKAR

Laxminarayan Institute of Technology, RTMNU, Amravati Road, Nagpur 440033, Maharashtra, India

Corresponding author. E-mail: adityathorgule@gmail.com

CONTENTS

ABSTRACT

Propionic acid is greatly utilized in food, chemical, pharmacy, and other industries. The gaining values of fermentation explained with new techniques leads to focus on technologies for downstream processing for product separation. The reactive extraction is effectively used to recover propionic acid from process feed solution. The reactive extraction with particular extractant and a specific pair of extractant and solvents give the higher capacity and selectivity of propionic acid. The extraction of propionic acid by reactive extraction is studied using solvents like toluene, n-hexane, petroleum ether (PE), and ethyl acetate (EA). Aqueous solution of propionic acid is taken between 0.1 and 0.4 gmol/L. The recovery of propionic acid (0.4 gmol/L) with solvents give the different distribution coefficient (K_D) in toluene, n-hexane, PE, and EA equal to 0.302, 0.126, 0.142, and 2.33, respectively. After the reactive extraction is studied with the combinations of extractant (tri-iso-octyl amine (TIOA))–solvents. The extraction of propionic acid with 30% of TIOA in solvent shows the higher K_D value for toluene, n-hexane, PE, and EA. These are 2.73, 0.843, 0.941, and 4.92, respectively. If results are compared then it can be concluded that using the combinations of extractant solvents, the extraction is effectively improved.

14.1 INTRODUCTION

Propionic acid is an important carboxylic acid basically used in food preservation, in the production of pesticides, polymers, and esters. It is also gaining its importance in pharmaceutical industry for synthesis of drugs, perfumes, and flavors. Due to such extensive applications, propionic acid attracts many researchers to find out different ways to increase its production commercially. Fermentation technology provides best alternative method to produce propionic acid. It requires best separation process to recover propionic acid from fermentation broth. Liquid extraction was employed with different organic solvents. Though extraction was successful, the amount of acid extracted was less which leads to find out advanced separation process. Reactive extraction was found to be an effective, ecofriendly, and economical technique to recover propionic acid.[1–5]

Extensive research has been done for reactive extraction of carboxylic acids with different organic solvents and extractant (complexing agent). Effect of primary to quaternary amines as extractant was studied. Ternary amines were found to be effective.[6–8] Wannesten studied the structure of amine–acid complex.[9] Effectiveness of extraction depends upon properties of organic solvents, extractant, operating parameters, and concentration of acid.[10–13] Some scientists tried to recover carboxylic acids from fermentation vessel and tried to optimize its recovery.[14–16] Yang studied the extraction of carboxylic acids at different pH values. Siebold noted that selection of the extractants should be based on the pH range of the feed solution of the acids, also on reusability of the extractants.[17] San Martin et al. and Yang et al. had concluded that the formation of the complex led the system to become more ordered and its entropy decreased. The concentration of acid extracted got decreased as the temperature increased.[18,19]

The objective of this work is to study the extraction of propionic acid with different organic solvents and in the presence of extractant (reactive extraction) by taking different initial concentration of propionic acid solution. The main focus of experimentation is to compare physical extraction with reactive extraction by calculating distribution in different organic solvents.

14.2 THEORY

Liquid–liquid extraction (LLE) is an important unit operation process in which feed mixture containing solute and solvent is brought in contact with another immiscible solvent. At equilibrium, solute distributes itself in solvents according to its relative solubility. The phase richer in solute is known as extract phase whereas another phase depleted in solute is known as raffinate phase. LLE is referred as solvent extraction process at many times.

Reactive extraction is a process of intensification method which involves extraction of solute with chemical reaction. In this process, the feed mixture is contacted with another phase containing a complexing agent which reacts reversibly with the solute of interest. Since the complex formation reaction can be selective for solutes with particular groups, these complexation reactions have the ability to separate particular solutes from feed. Reactive extraction is an economical process because organic solvents can be completely recovered, further can be reutilized, along with cheap and simple separation techniques.

14.3 EXPERIMENTAL PROCEDURE

A rigorous experimentation is carried out to extract propionic acid from the aqueous solution using physical as well as reactive extraction method in this work. The experiment was performed at the atmospheric pressure of 27°C (room temperature). Physical extraction was carried out by taking *n*-hexane, toluene, petroleum ether (PE), and ethyl acetate (EA) as an organic solvent. As fermentation produces fewer amounts of carboxylic acids, the initial concentration of propionic acid was kept in between 0.1 and 0.4 gmol/L. Equal volume of propionic acid solutions and pure organic solvents were taken. The mixture was shaken in the shaker incubator for 5 h followed by settling for 2 h into the separating funnel. Organic phase got separated by aqueous phase. Analysis was done by titrating aqueous phase sample with standard NaOH solution (0.025 N). Burette readings were taken when pink color appeared.

Organic phase concentration was calculated by mass balance equation. Further, distribution coefficient for each case was calculated and *n* data was tabulated. For reactive extraction, 20 and 30% by volume tri-iso-octyl amine (TIOA) as an extractant was added in pure organic solvent and similar procedure was repeated by taking initial concentration of propionic acid solution as 0.4 gmol/L.

14.4 RESULTS AND DISCUSSIONS

Distribution of propionic acid in both aqueous and organic phases was calculated and tabulated. Data in Table 14.1 show that the propionic acid got extracted by organic solvents. It is distributed differently in both phases for different initial concentration of aqueous solution. Propionic acid distribution in both phases is shown graphically in Figure 14.1. It shows that as propionic acid concentration in organic phase increases, the concentration in the aqueous phase also increases.

TABLE 14.1 PA Distribution in Solvents (Physical Extraction).

Initial concentration [A] gmol/L	Toluene		n-Hexane		Petroleum ether		Ethyl acetate	
	[HA] aq	[HA] org	[HA] aq	[HA] org	[HA] aq	[HA] org	[HA] aq	[HA] org
0.1	0.085	0.015	0.088	0.012	0.095	0.005	0.037	0.063
0.2	0.175	0.025	0.189	0.011	0.185	0.015	0.065	0.135
0.3	0.232	0.0675	0.270	0.030	0.264	0.036	0.092	0.208
0.4	0.307	0.093	0.355	0.045	0.350	0.050	0.120	0.280

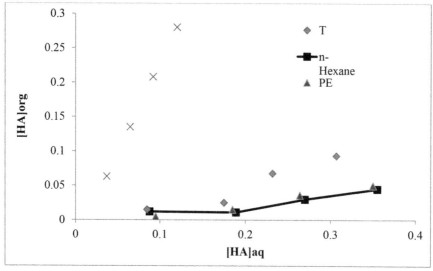

FIGURE 14.1   [HA]org vs [HA]aq.

The tabulated data show different distribution pattern of propionic acid in different solvents. EA was found as the best solvent in this study. Propionic acid reacts more with the solvents possessing higher dipole moments. Thus, the distribution of propionic acid given in table is justified (Table 14.2).

TABLE 14.2   Distribution Coefficient in Different Solvents for Physical Extraction.

Initial concentration [A] gmol/L	Distribution coefficient in different solvents K_D			
	Toluene	Hexane	PE	EA
0.1	0.176	0.129	0.0526	1.66
0.2	0.142	0.055	0.081	2.07
0.3	0.276	0.108	0.136	2.24
0.4	0.302	0.126	0.142	2.33

Graphical representation (Fig. 14.2) of distribution coefficient K_D with initial concentration of aqueous solution of propionic acid [A] shows increase in value of K_D along with increase in value of [A] for PE and EA whereas decrease and further increase in the distribution coefficient for toluene and n-hexane.

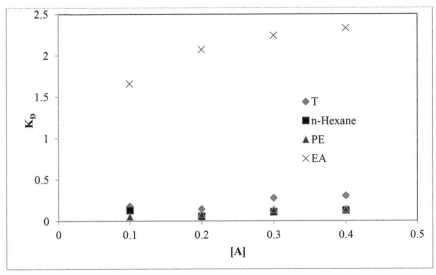

FIGURE 14.2 K_D vs [A].

For reactive extraction, distribution of propionic acid in both phases as well as distribution coefficient in each solvent was calculated and tabulated (Tables 14.3 and 14.4).

TABLE 14.3 Distribution of PA in Different Solvents (Reactive Extraction).

Extractant % by volume in solvents	Toluene		n-Hexane		Petroleum ether		Ethyl acetate	
Initial concentration (0.4 gmol/L)	[HA] aq	[HA] org	[HA] aq	[HA] org	[HA] aq	[HA] org	[HA] aq	[HA] org
20	0.134	0.266	0.255	0.145	0.250	0.150	0.071	0.329
30	0.107	0.293	0.217	0.183	0.206	0.194	0.067	0.332

TABLE 14.4 Distribution Coefficient in Different Solvents for Reactive Extraction.

Extractant % by volume in solvents	Distribution Coefficient in Different Solvents K_D			
Aqueous solution (0.4 gmol/L)	Toluene	*n*-Hexane	Petroleum ether	Ethyl acetate
20	1.985	0.560	0.600	4.63
30	2.730	0.843	0.941	4.92

Graphical representation of distribution coefficient in different organic solvents with volume % of TIOA in solvents shows a large increase in the values of distribution coefficient with increase in volume % of TIOA in organic solvents (Fig. 14.3). Hence, we can say that reactive extraction is more favorable than that of physical extraction.

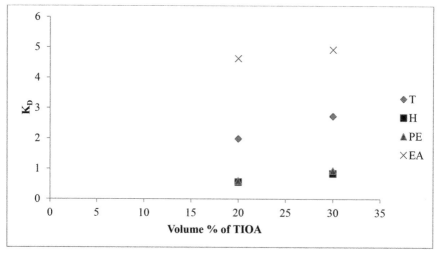

FIGURE 14.3 K_D vs volume % of TIOA.

14.5   CONCLUSION

Propionic acid was extracted successfully by reactive extraction method using TIOA as extractant with different organic solvents. Propionic acid gave different concentration distribution pattern in both phases according to the initial concentration of it in aqueous solution. The distribution coefficient values of propionic acid while extracted by only solvent had lower values of distribution coefficients whereas increased values of distribution coefficient were obtained by reactive extraction. EA was found to be the best organic solvent to extract propionic acid. Enhanced increase in the K_D, that is, distribution coefficients (200–215%) were obtained with increase in % volume of extractant in organic solvents.

KEYWORDS

- **propionic acid**
- **solvents**
- **reactive extraction**
- K_D
- **TIOA**

REFERENCES

1. Keshav, A.; Chand, S.; Wasewar, K. L. Recovery of Propionic Acid from Aqueous Phase by Reactive Extraction Using Quarternary Amine (Aliquat 336) in Various Diluents. *Chem. Eng. J.* **2009**, *152*, 95–102.
2. Wasewar, K. L.; Yoo, C. K. In *Intensifying the Recovery of Carboxylic Acids by Reactive Extraction*, 3rd International Conference on Chemistry and Chemical Engineering, Jeju, Island, South Korea, Jun 29–30, 2012; IPCBEE: Singapore, 2012.
3. Keshav, A.; Wasewar, K. L.; Chand, S. Extraction of Propionic Acid with Tri-n-octyl Amine in Different Diluents. *Sep. Purif. Technol.* **2008**, *63*, 179–183
4. Wasewar, K. L. Reactive Extraction: An Intensifying Approach for Carboxylic Acid Separation. *Int. J. Chem. Eng. Appl.* **2012**, *3*, 249.
5. Uslu, H.; Inci, I. (Liquid + liquid) Equilibria of the (Water + Propionic Acid + Aliquat 336 + Organic Solvents) at T = 298.15 K. *J. Chem. Thermodyn.* **2007**, *39*, 804–809.
6. Wasewar, K. L.; Pangarkar, V. G. Intensification of Propionic Acid Production by Reactive Extraction: Effect of Diluents on Equilibrium. *Chem. Biochem. Eng. Q.* **2006**, *20*, 325–331.
7. Gu, Z.; Glatz, B.; Glatz, C. E. Propionic Acid Production by Extractive Fermentation. I. Solvent Considerations. *Biotechnol. Bioeng.* **1998**, *57*, 454–461.
8. Kumar, S.; Babu, B. V. Extraction of Pyridine-3-Carboxylic Acid Using 1-Di Octylphosphoryloctane (TOPO) with Different Diluents: Equilibrium Studies. *J. Chem. Eng. Data.* **2009**, *54*, 2669–2677.
9. Wennersten, R. Extraction of Carboxylic Acid from Fermentation Broth Using Solution of Tertiary Amine. *J. Chem. Technol. Biotechnol.* **1983**, *33-B*, 85–94.
10. Ingale, M. N.; Mahajani, V. V. Recovery of Carboxylic Acids, C2-C6, from an Stream Using TBP: Effect Aqueous Waste of Presence of Inorganic Acids and their Sodium Salts. *S.E.P. Technol.* **1996**, *6*, 1–7.
11. Morales, A. F.; Albet, J.; Kyuchoukov, G.; Malmary, G.; Molinier, J. Influence of Extractant (TBP and TOA), Diluent, and Modifier on Extraction Equilibrium of Monocarboxylic Acids. *J. Chem. Eng. Data.* **2003**, *48*, 874–886.
12. Gu, Z.; Glatz, B.; Glatz, C.E. Propionic Acid Production by Extractive Fermentation I. Solvent Considerations. *Biotechnol. Bioeng.* **1998**, *57*, 454–461.
13. Matsumoto, M.; Otono, T.; Kondo, K. Synergistic Extraction of Organic Acids with TOA and TBP. *Sep. Purif. Technol.* **2001**, *24*, 337–342.

14. Juang, R.S.; Huang, R.H. Equilibrium Studies on Reactive Extraction of Lactic Acid with an Amine Extractant. *Chem. Eng. J.* **1997,** *65,* 47–53.
15. Wasewar, K. L.; Heesink, A. B. M.; Versteeg, G. F.; Pangarkar, V. G. Equilibria and Kinetics for Reactive Extraction of Lactic Acid Using Alamine 336 in Decanol. *J. Chem. Technol. Biotechnol.* **2002,** *77,* 1068–1075.
16. Poposka, A. F.; Nikolovski, K.; Tomovska, R. Kinetics, Mechanism and Mathematical Modeling of Extraction of Citric Acid with Isodecanol/N-Paraffins Solutions of Trioctylamine. *Chem. Eng. Sci.* **1998,** *53,* 3227–3237.
17. Siebold, M.; Frieling, P. V.; Joppien, R.; Rindfleisch, D.; Schügerl, K.; Röper, H. Comparison of the Production of Lactic Acid by Three Different Lactobacilli and its Recovery by Extraction and Electrodialysis. *Process Biochem.* **1995,** *30,* 81–95.
18. Yang, S.; White, S. A.; Hsu, S. Extraction of Carboxylic Acids with Tertiary and Quaternary Amines: Effect of pH. *Ind. Eng. Chem. Res.* **1991,** *30,* 1335–1342.
19. San-Martin, M.; Pazos, C.; Coca, J. Reactive Extraction of Lactic Acid with Alamine 336 in the Presence of Salts and Lactose. *J. Chem. Technol. Biotechnol.* **1996,** *65,* 1–6.

PART III
Modeling and Simulation of Chemical Processes

CHAPTER 15

FLOW CHARACTERISTICS OF NOVEL SOLID–LIQUID MULTISTAGE CIRCULATING FLUIDIZED BED

M. A. THOMBARE and P. V. CHAVAN[*]

Department of Chemical Engineering, Bharati Vidyapeeth Deemed University, College of Engineering, Pune 411043, Maharashtra, India

[*]*Corresponding author. E-mail: pvcuict@gmail.com; pvchavan@ bvucoep.edu.in*

CONTENTS

ABSTRACT

A multistage solid–liquid circulating fluidized bed (SLCFB) was constructed to accommodate both the steps of operation of loading (catalytic reaction, adsorption, etc.) and regeneration of solid phase simultaneously on a continuous mode. The loading and regeneration sections primarily consists of five glass stages (each 100 mm i.d. and 100 mm height) assembled together with flange joint. A stainless steel (SS) sieve plate of 5% opening and 2 mm hole size is fixed between the flanges. Each SS plate comprises at least one downspout either located at the center or at the periphery of the stage, imparting the flow behavior of solid close to plug flow. The polymeric resins were used as solid phase and water as liquid phase, respectively. In this work, we have studied the effect of system (particle density and size of solid phase) and operating (superficial liquid velocity and solid circulation rate) parameters on pressure drop across the section. The results indicate that a modified cross flow stage configuration imparts a uniform fluidization, which in turn results into reduction in pressure drop as compared to SLCFBs studied so far.

15.1 INTRODUCTION

Solid–liquid fluidized beds (SLFBs) are commonly used in chemical, petrochemical, pharmaceutical, and metallurgical industries as classifier, crystallizer, solid–liquid reactor, and expanded bed chromatography, etc. The fixed bed mode of operation is, however, associated with four major limitations: (a) batch-wise overall operation, (b) high pressure drop, (c) clogging of bed if feed contains suspended solids, and (d) restriction of utilization of bed to the small fraction of total bed height, called as active zone. These constraints make overall operation time and energy intensive. Expanded bed mode of operation overcomes the constraints of fixed bed related to high pressure drop and clogging of bed. However, it is necessary to address disadvantages of fixed bed to increase the fraction of total bed height as an active zone and throughput.

Solid–liquid circulating fluidized bed (SLCFB) overcomes the constraints associated with fixed bed and expanded bed. It mainly comprises of a riser column and a downcomer. The riser column is operated in a circulating fluidization regime while the downcomer is operated in fixed bed or expanded bed modes. The liquid streams to the riser column and downcomer are supplied separately without intermixing. The solid and liquid phases are contacted co-currently and counter-currently in the riser and downcomer, respectively.

SLCFBs have come up with many outstanding features such as continuous circulation of solid phase for higher throughputs, efficient mass transfer, and reduced back-mixing. These distinctive attributes of SLCFB make them appropriate for various industrial applications such as manufacture of linear alkylbenzene,[1,2] fermentation products' recovery in continuous mode,[3-5] treatment of radioactive nuclear liquid waste streams to remove and recover heavy metal,[6] continuous removal of dissolved ions from wastewater,[7] and enzymatic polymerization reaction of phenol in bio-refining process.[8]

SLCFBs studied so far, however, demands an appropriate pressure balance and a proper dynamic seal connecting the two sections for stable and satisfactory operation. Further, higher solid phase and liquid phase mixing is expected in the riser column since it operates in circulating fluidization regime. The higher liquid velocity in the riser column (more than particle terminal settling velocity) may lead to huge height if loading and/or regeneration dynamics are time intensive. It would be worthwhile to mention that geometrical configuration of the multistage SLCFBs studied so far resemble closely to conventional distillation column wherein stages are structured with eccentric downspout leading to non-uniform flow of solid particles. The non- ideality of flow pattern with respect to the solid particles causes decline in overall mass transfer rate. Therefore, it is essential to construct a new stage configuration to supersede the limitation of existing configuration of SLCFBs.

In order to design and develop a proposed SLCFB, it is essential to study flow structure existing in the SLCFB since it governs the design parameters such as heat transfer and mass transfer coefficient, dispersion coefficient, etc. It has long been recognized that the radial and axial flow structure prevailing in solid–liquid fluidization is uniform in conventional fluidization regime. This homogeneous behavior forms the basis of theoretical background of conventional solid–liquid systems.[9-11] Experimental results also confirm that conventional solid–liquid fluidization is indeed homogeneous.[9,12-17] The flow pattern, however, becomes non-uniform when superficial liquid velocity exceeds particle terminal velocity.[18-27] It is clear that the hydrodynamic behavior of SLCFB differs greatly as compared to SLFB, mainly in context to flow pattern, and liquid and solid hold up. The uniform flow pattern in SLFB abruptly becomes non-uniform in SLCFB in axial as well as radial directions. Moreover, the flow pattern and hold-up profile are greatly affected by superficial liquid velocity, physical properties of solid particles and liquid phase, and solid circulation rate.[2,21,23,28-31] Therefore, this work concentrates on the effect of solid phase physical properties and operating parameters on the performance of novel SLCFB, originally proposed by Lali et al.[32]

15.2 EXPERIMENTAL SECTION

15.2.1 MATERIALS

The solid phase was used T-42 (Thermax made) strong acid cation exchange resin. Table 15.1 shows the properties of the resin. Water was used as liquid phase.

TABLE 15.1 Resin Properties as given by the Manufacturer.

Property	TULSION T-42
Type	Strong acid cation exchange resin
Particle size (mm)	0.3–1.2
Total exchange capacity (meq/mL resin)	1.8
Moisture content (% kg water/kg wet resin)	52 ± 3
Functional group	$-SO_3^-$
Ionic form	H^+

15.2.2 EXPERIMENTAL SETUP

The experimental multistage solid–liquid circulating fluidization system is as shown in Figure 15.1. It has two main sections (a) loading section and (b) regeneration section. Each section consists of five glass stages (each 100 mm i. d. and 100 mm height) assembled together with flanges. A *stainless steel* (SS) sieve plate of 5% opening and 2 mm hole size is fixed between the flanges. A SS mesh of hole size smaller than the solid particle is fitted on the SS plate. The sieve plate has two designs namely, one having a downspout at the center and the other having two downspouts at the diagonal ends of the periphery. These sieve plates are alternately arranged between the glass stages. The solid particles move on the underneath SS sieve plate through the downspout. The liquid moves counter-currently upwards through the holes of SS sieve plate.

The inlets are at the bottom of each section which comprises of primary and auxiliary inlets, and outlet is at the top of each section. The primary flow is introduced through a conical distributor of 70–75 mm diameter covered with a SS mesh. The auxiliary inlet is located below primary inlet of each section and introduced through the SS mesh. The arrangement of primary inlet and auxiliary inlet of both sections are similar. An arrangement is made

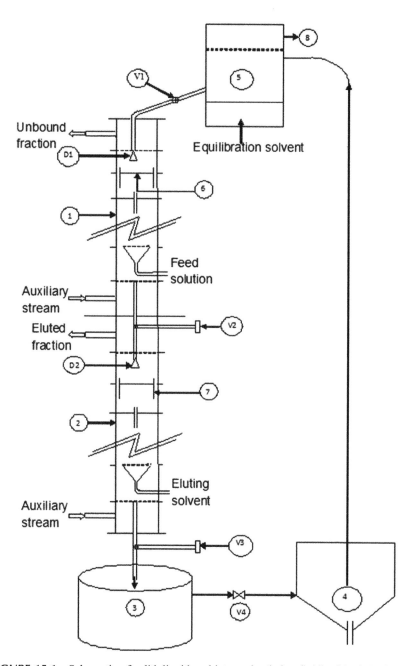

FIGURE 15.1 Schematic of solid–liquid multistage circulating fluidized bed: 1—loading section, 2—regenerating section, 3- Solid particles accumulating tank, 4-Solid liquid ejector, 5-Solid feed tank, 6-SS stage fitted with SS mesh 7-Downcomer, 8-Overflow liquid, V. Valve; D. Diffuser/solid distributor (All dimensions are in mm).

for the measurement of the pressure drop across the section using a U-tube manometer. The inlet liquid flow rate is measured through rotameter and controlled by ball valves.

The regenerated solid particles are charged from a feed tank (350 mm i. d. and 500 mm length) through valve V1 to the loading section at the top. A diffuser in the loading section distributes the solid particles evenly on the sieve plate. The feed tank is provided with the equilibration liquid through a calming section that keeps the solid particles in expanded state. A top outlet is provided to collect the equilibration liquid through a SS mesh so as to prevent drainage of fine solid particles. The regenerated solid particles flow from the loading section to the regenerating section through a valve connected between them. A plug of fixed bed is maintained for the smooth and continuous movement of the solid particles. Solid particles are distributed uniformly through a diffuser on the sieve plate of the regenerating section. The solid particles move downwards through the downspout of the sieve plate to the next stage. The solid particles at the bottom section are continuously fed to the feed tank through solid conveying pipe. Thus, the circulation of solid particles takes place continuously.

15.2.3   EXPERIMENTAL PROCEDURE

The resin particles were sieved into two sizes namely 0.6 and 0.4 mm, prior to the experiments carried out on the multistage SLCFB. The swelling of the resin particles was determined by measuring initial volume of the dry resin particles and final volume of the swollen resin particles (after sorption experiment) using standard calibrated glass tubes.

A predetermined amount of the resin (1.5 kg) of a known size was put into the solid feed tank. The liquid flow was started and metered using rotameters. Ball valves were used to control the liquid flow. The solid particles moved from the loading section into the regenerating section through the valve V2. A fixed bed is maintained in the region above the valve V2, so that there is smooth flow of solid in the regenerating section and no liquid diffuses into the loading section. After reaching steady state, the pressure drop measurements were performed in the loading and regenerating section by U-tube manometer. The manometric fluid used was bromo-benzene having specific gravity 1.45. The solid flow rate was measured through a three-way valve.

15.3 RESULTS AND DISCUSSION

15.3.1 SWELLING CHARACTERISTICS OF THE RESIN

The polymeric structure of the resin swells significantly when brought in contact with water. The swelling ratio (S_R) is defined as the ratio of volume of swollen particles to the volume of dry particles. Table 15.2 shows the values of swelling characteristics of the resin. The solvation of fixed ionic groups and counter-ions of the resin mainly contribute to the swelling of the resin because of their ability to form ion-dipole bonds. The positive pole of water is attracted toward nuclear sulfonic group (SO_3^-) of the resin whereas negative pole toward the counter-ion (H^+). Therefore, around each fixed charge and respective counter-ion, a cluster of water molecules is formed. This, in turn, swells the resin significantly.

As seen in Table 15.2, the change in the average resin size of 0.6 and 0.4 mm particles after swelling are 13.33 and 10.0%, respectively. It is worthwhile to note that particle terminal settling velocity is strongly affected by the particle diameter. The particle terminal settling velocity increased with the increase of particle diameter. This, in turn, changes the flow field around the particle and therefore, it is essential to consider the swelling of the particle.

15.3.2 EXPANSION CHARACTERISTICS

Experimental work was carried out in an acrylic column of 100 mm i. d. and 1.2 m length to study expansion characteristics of the solid particle of a given size. The calming section, 300 mm long, was filled with sand particles and positioned at the bottom of the column to minimize the entrance energy losses. The water flows through distributor plate, made up of acrylic material having 580 holes of 2 mm diameter in square pitch (fractional opening is 23.2%). A SS mesh of opening size smaller than particle size was fitted to the distributor plate to avoid the movement of particles through the distributor plate. The pressure taps were introduced at 50 mm above distributor plate at an interval of 100 mm over the column length. Pressure difference was measured using U-tube manometer with nitrobenzene having specific gravity of 1.20.

The voidage of bed with respect to superficial liquid velocity has been measured by two methods namely: (a) solid-phase balance method and (b) pressure gradient method.

TABLE 15.2 Swelling Characteristics of Resin.

Particle size (mm)	Volume of particle (mL)		Swelling ratio	Diameter of particle (mm)		Galileo number (Ga)		Rise in particle diameter
	Dry basis	Wet basis		Dry basis	Wet basis	Dry basis	Wet basis	
0.6	6	8.9	1.48	0.6	0.68	635.68	925.37	1.13
0.4	6.05	8.1	1.39	0.4	0.44	188.35	250.69	1.10

In solid-phase balance method, the average voidage is calculated as follows:

$$\epsilon_S = \frac{\epsilon_{SO} H_o}{H}$$ (15.1)

In pressure gradient method, the average voidage is calculated as follows by neglecting wall effects and solid accelerations,

$$-\left(\frac{dP}{dz}\right)_z = (1-\epsilon_{LZ})(\rho_S - \rho_L)g$$ (15.2)

Figure 15.2 shows the effect of superficial liquid velocity on bed voidage by using solid balance and pressure gradient method. It has been observed that both the methods measure bed voidage with a standard deviation of ±2.07%.

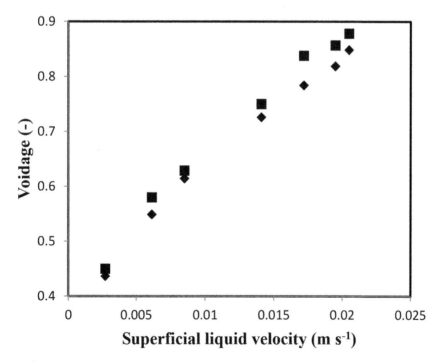

FIGURE 15.2 Bed voidage variation at various superficial liquid velocities along the bed length for 0.6 mm particle size: (■) pressure gradient method and (♦) solid-phase balance method.

The Richardson–Zaki parameter has been estimated by plotting $\ln V_L$ vs $\ln \varepsilon_L$. The data of Richardson–Zaki parameters for particles are reported in Table 15.3.

TABLE 15.3   Richardson–Zaki Index.

Particle size (mm)	Particle terminal settling velocity, Vs_∞ (mm s^{-1})	Richardson–Zaki index, n (−)
0.6	36.11	3.31
0.4	29.15	3.50

15.3.3   PRESSURE DROP ACROSS THE MULTISTAGE COLUMN

The pressure drop across the multistage column depends on system (physical properties of solid phase and liquid phase), operating (superficial liquid velocity and solid circulation rate), and geometrical parameters (internals of column). So as to enumerate the effect of these parameters, experiments were carried out when: (a) there is no solid particle flow from one to another stage, and (b) SLCFB is operating at a steady state. The hydrostatic pressure drop measured theoretically (equal to hρg, where h is the height of the column) was 7.06 kPa. The contribution of wall friction at the column wall to the pressure drop is considered to be negligible.

15.3.3.1   PRESSURE DROP WITHOUT SOLID FLOW

In order to delineate the effect of solid particle mass and size on the pressure drop, it was measured across fixed and fluidized beds of solid particles for different mass and particle sizes. Figure 15.3 shows pressure drop as a function of superficial liquid velocity for 0.6 and 0.4 mm particle sizes. The experimental data are shown for the solid particles filled up to 20 and 12 mm on each stage, corresponding to 0.63 and 0.38 kg of resin, respectively, for two particle sizes namely 0.6 and 0.4 mm. The pressure drop increases as the amount of resin on each stage increases. This is obvious because pressure drop depends on the actual weight of the solid particles. The minimum fluidization velocity has been estimated theoretically by equating eq 15.2 with Ergun equation. Figure 15.3 gives the experimental value of minimum fluidization velocity for 0.6 mm particle size as 2.17 mm s^{-1} while theoretical value obtained is 2.16 mm s^{-1}. The theoretical and experimental values of minimum fluidization velocity for particle size of 0.4 mm are 0.98 and 1.03 mm s^{-1} correspondingly. The theoretical and experimental values match well. It is also observed from Figure 15.3 that theoretical curves (solid lines in Fig. 15.3) and the data for fixed bed condition were found to be closely foreseen with an average standard deviation of 1.6%.

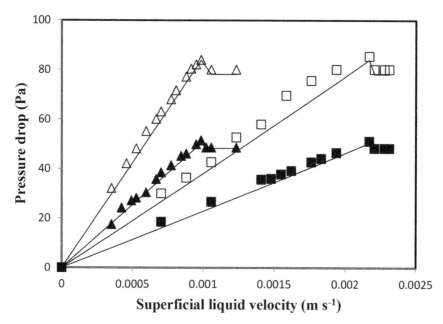

FIGURE 15.3 Pressure drop in fixed bed of resin for varying particle size and bed height per stage ((□): 0.6 mm particle size and 20 mm bed height, (Δ): 0.4 mm particle size and 20 mm bed height, (■): 0.6 mm particle size and 12 mm bed height, (▲): 0.4 mm particle size and 12 mm bed height, and (___): theoretical values).

15.3.3.2 *PRESSURE DROP WITH SOLID CIRCULATION*

Experiments have also been performed at steady flow of solid particles for a known size with superficial liquid velocity ranging from minimum fluidization velocity to terminal settling velocity of solid particles. The solid circulation rate was kept constant at 6 g s^{-1}. It has been experimentally seen that water mainly flows through the mesh (>90%) and small fraction flows through downspout since ratio of diameter of column to diameter of downspout >12.5. Figure 15.4 shows that pressure drop increases with the increase of superficial liquid velocity. When compared to Figure 15.3, it is evident that pressure drop due to the counter-current interaction between solid particles and water is negligible when compared with pressure drop offered by internals and apparent weight of solid phase. It would be worthwhile to mention that hydrostatic head mainly contributes to the pressure drop along the column.

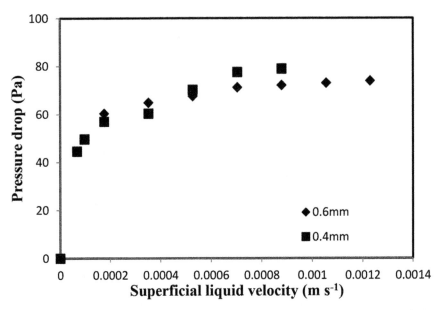

FIGURE 15.4 Pressure drop in the regenerating column at total superficial liquid velocity at fixed primary flow rate and constant solid flow rate of 6 g/s for particle size of (♦: 0.6 mm and ■: 0.4 mm).

15.4   CONCLUSION

From this study, conclusions put forth are as follows:

1. Swelling of the resin particles should be considered while estimating Richardson–Zaki parameters since swelling of the resin particles significantly increases the size of the resin particles. In the current case, the swelling ratio of the resin particles was found to be 1.48 and 1.38 for 0.6 and 0.4 mm particle sizes, respectively.
2. In multistage SLCFB hydrostatic head mainly contributes (95%) to the overall pressure drop across the column.

ACKNOWLEDGMENTS

The authors acknowledge financial support from the Department of Science and Technology (DST), Govt. of India, New Delhi (DST No.: SB/S3/CE/025/2014-SERB Dated 14/7/2014) to carry out this study.

NOMENCLATURE

H = solid bed height, m
P = pressure drop in the segment under consideration, N m^{-2}
Z = height of column from bottom to point of consideration, m
g = gravitational acceleration, m s^{-2}
h = height of column, m

Greek letters
ε = voidage of the bed
ρ = density, kg m^{-3}

Subscripts
o = initial condition
L = liquid
S = solid

KEYWORDS

- solid–liquid circulating fluidized bed
- liquid fluidization
- flow characteristics
- hydrodynamics

REFERENCES

1. Liang, W. G.; Yu, Z.; Jin, Y.; Wang, Z; Wang, Y. Synthesis of Linear Alkyl Benzenes in a Liquid-Solid Circulating Fluidized Bed Reactor. *J. Chem Technol. Biotechnol.* **1995,** *62,* 98–102.

2. Liang, W. G.; Zhu, J. X. Effect of Radial Flow Non-Uniformity on the Alkylation Reaction in a Liquid-Solid Circulating Fluidized-Bed (LSCFB) Reactor. *Ind. Eng. Chem. Res.* **1997**, *36*, 4651–4658.

3. Lan, Q.; Bassi, A. S.; Zhu, J. X.; Margitis, A. Continuous Protein Recovery with a Liquid-Solid Circulating Fluidized-Bed Ion Exchanger. *AIChE. J.* **2002**, *48*, 252–261.

4. Lan, Q.; Bassi, A. S.; Zhu, J. X.; Margitis, A. Continuous Protein Recovery from whey using Liquid-Solid Circulating Fluidized-Bed Ion Exchange Extraction. *Biotechnol. Bioeng.* **2002**, *78*, 157–163.

5. Lan, Q.; Bassi, A. S.; Zhu, J. X.; Margitis, A. Continuous Protein Recovery using Liquid-Solid Circulating Fluidized-Bed Ion Exchange Extraction System: Modelling and Experimental Studies. *Can. J. Chem. Eng.* **2000**, *78*, 858–866.

6. Feng, X.; Jing, S.; Wu, Q.; Chen, J.; Song, C. The Hydrodynamic Behaviour of the Liquid-Solid Circulating Fluidized Bed Ion Exchange for Cesium Removal. *Powder Technol.* **2003**, *134*, 235–242.

7. Kishore, K.; Verma, N. Mass Transfer Studies in Multi-Stage Counter Current Fluidized Bed Ion Exchangers. *Chem. Eng. Process.* **2006**, *45*, 31–35.

8. Trivedi, U.; Bassi, A. S.; Zhu, J. X. Continuous Enzymatic Polymerization of Phenol in a Liquid-Solid Circulating Fluidized Bed. *Powder Technol.* **2006**, *169*, 61–70.

9. Richardson J. F.; Zaki, W. N. Sedimentation and Fluidization. *Trans. Inst. Chem. Eng. Part A.* **1954**, *32*, 35–53.

10. Kwauk, M. Generalized Fluidization, I. Steady-State Motion. *Scientia Sinica.* **1963a**, *12* (4), 587–612.

11. Kwauk, M. Generalized Fluidization, II. Accelerative Motion with Steady Profiles. *Scientia Sinica.* **1964a**, *13* (9), 1477–1492.

12. Wilhem, R. H.; Kwauk, M. Fluidization of Solid Particles. *Chem. Eng. Prog.* **1948**, *44*, 201–217.

13. Mertes, T. S.; Rhodes, H. B. Liquid-Particle Behavior. *Chem. Eng. Prog.* **1955**, *51*, 429–432, 517–522.

14. Lapidus, L.; Elgin, J. C. Mechanics of Vertical-Moving Fluidized Systems. *AIChE. J.* **1957**, *3*, 63–68.

15. Foscolo, P. U.; Gibilaro, L. G. Fluid Dynamic Stability of Fluidized Suspensions: The Particle Bed Model. *Chem. Eng. Sci.* **1987**, *42*, 1489–1500.

16. Khan, A. R.; Richardson, J. F. Fluid-Particle Interactions and Flow Characteristics of Fluidized Beds and Settling Suspensions of Spherical Particles. *Chem. Eng. Commun.* **1989**, *78*, 111–130.

17. Chavan, P. V.; Joshi, J. B. Analysis of Particle Segregation and Intermixing in Solid-Liquid Fluidized Bed. *Ind. Eng. Chem. Res.* **2008**, *47* (21), 8458–8470.

18. Liang, W. G.; Zhu, J. X.; Jin, Y.; Yu, Z. Q.; Wang, Z. W.; Zhou, J. Radial Non-Uniformity of Flow Structure in a Liquid-Solid Circulating Fluidized Bed. *Chem. Eng. Sci.* **1996**, *51*, 2001–2010.

19. Liang, W. G.; Zhang, S. L.; Zhu, J. X.; Jin, Y.; Yu, Z. Q.; Wang, Z. W. Flow Characteristics of the Liquid-Solid Circulating Fluidized Bed. *Powder Technol.* **1997a**, *90*, 95–102.

20. Liang, W. G.; Zhu, J. X. A Core-Annulus Model for the Radial Flow Structure in a Liquid-Solid Circulating Fluidized Bed (LSCFB). *Chem. Eng. J.* **1997b**, *68*, 51–62.

21. Zheng, Y; Zhu, J. X.; Wen, J; Martin, S.; Bassi, A.; Margaritis, A. The Axial Hydrodynamic Behavior in a Liquid-Solid Circulating Fluidized Bed. *Can. J. Chem. Eng.* **1999**, *77*, 284–290.

22. Roy, S.; Dudukovic, M. P. Flow Mapping and Modeling of Liquid-Solid Risers. *Ind. Eng. Chem. Res.* **2001**, *40*, 5440–5454.

23. Zheng, Y.; Zhu, J. X. The Onset Velocity of a Liquid Solid Circulating Fluidized Bed. *Powder Technol.* **2001**, *114*, 244–251.

24. Razzak, S. A.; Agarwal, K.; Zhu, J.; Zhang, C. Numerical Investigation on the Hydrodynamics of an LSCFB Riser. *Powder Technol.* **2008**, *188* (1), 42–51.

25. Cheng, Y.; Zhu, J. Hydrodynamics and Scale-Up of Liquid-Solid Circulating Fluidized Beds: Similitude Method vs. CFD. *Chem. Eng. Sci.* **2008**, *63* (12), 3201–3211.

26. Vidyasagar, S.; Krishnaiah, K.; Sai, P. S. T. Macroscopic Properties of Liquid-Solid Circulating Fluidized Bed with Viscous Liquid Medium. *Chem. Eng. Process. Process Intensif.* **2011**, *50* (1), 42–52.

27. Razzak, S. A.; Rahman, S. M.; Hossain, M. M.; Zhu, J. Artificial Neural Network and Neuro-Fuzzy Methodology for Phase Distribution Modelling of a Liquid-Solid Circulating Fluidized Bed Riser. *Indus. Eng. Chem. Res.* **2012**, *51*, 12497–12508.

28. Zheng, Y.; Zhu, J. X.; Marwaha, N.; Bassi, A. S. Radial Solid Flow Structure in a Liquid-Solid Circulating Fluidized Bed. *Chem. Eng. J.* **2002**, *88*, 141–150.

29. Chavan, P. V.; Kalaga, D. V.; Joshi, J. B. Solid-Liquid Circulating Multistage Fluidized Bed: Hydrodynamic Study. *Ind. Eng. Chem. Res.* **2009**, *48*, 4592–4602.

30. Sang, L.; Zhu, J. Experimental Investigation of the Effects of Particle Properties on Solids Holdup in an LSCFB Riser. *Chem. Eng. J.* **2012**, *197*, 322–329.

31. Dadashi, A.; Zhang, C.; Zhu, J. Numerical Simulation of Counter-Current Flow Field in the Downcomer of a Liquid-Solid Circulating Fluidized Bed. *Particuology.* **2015**, *21*, 48–54.

32. Lali, A. M.; Kale, S. B.; Pakhale, P. D.; Thakare, Y. N. Continuous Countercurrent Fluidized Moving Bed (FMB) and/or Expanded Moving Bed (EMB). U.S. Patent 8,673,225, 2014.

CHAPTER 16

MATHEMATICAL MODELING AND SIMULATION OF GASKETED PLATE HEAT EXCHANGER

J. KATIYAR[1] and S. HUSAIN[2,*]

[1]*Department of Chemical Engineering, Harcourt Butler Technological Institute, Kanpur 208002, Uttar Pradesh, India*

[2]*Department of Chemical Engineering, Aligarh Muslim University, Aligarh 202002, Uttar Pradesh, India*

Corresponding author. E-mail: drsattarhusain@yahoo.co.in

CONTENTS

ABSTRACT

This study presents a mathematical model which is developed for the simulation of gasketed plate heat exchanger (PHE) with general configuration and operating at steady state. The purpose of developing this type of model is to study the influence of the configuration on the exchanger performance for rigorous configuration selection. A mathematical model of PHE in steady state yields a linear system of first order ODE comprising of the energy balance for each channel and required boundary condition. The objective is to make a numerical simulation program which acts as a tool to examine the combined effect of most of the geometrical parameter on the operation of the PHE and to check to what extent, each of these parameters affects the momentum and heat transfer rate. In this effort, the simulation work is based on RK-4 numerical approach for the system of ODE-45 and finite difference method numerical approach, developed in MATLAB. For the entire study, the results show the little influence over the main simulation result over the heat exchanger (thermal effectiveness, pressure drop, and outlet temperature of process fluid) as far as overall heat transfer coefficient (U) is concerned.

16.1 INTRODUCTION

A plate heat exchanger (PHE) or plate and frame heat exchanger is a device constructed for efficient heat transfer between two mediums (fluid streams); that is, one hot stream and another cold stream are transported into "thermal contact" in arrangement to transfer the heat from the hot stream to cold stream. Gasketed PHEs are effective in heat transfer, easy to clean, adjustable, and environmentally effective, and are widely used in chemical and food processing industries, space heating, condensation, evaporation, and many other industries including oil and gas industry.[1-13]

16.2 NOMENCLATURE

a general PHE model parameter	G_p port mass velocity, kg/m²s	R fluid fouling factor, m²°C/W	α dimensionless heat transfer parameter
A effective plate heat transfer area, m²	h convective heat transfer coefficient, W/m²°C	Re Reynolds number, $Re = G_c De/\mu$	β chevron corrugation inclination angle, degrees

b channel average thickness, m	k fluid thermal conductivity, W/m°C	S_i channel i flow direction binary parameter, $S_i = +1$ or -1	I the i-th element
C_p fluid specific heat, J/ kg°C	K_p plate thermal conductivity, W/m°C	T temperature,°C	ΔP fluid pressure drop, Pa
D_e channel equivalent diameter, m	L effective length of plate, m	U heat transfer coefficient (overall), W/m²°C	ε_p plate thickness, m
D_p port diameter of plate, m	N number of channels per pass	w effective width of plate, m	ρ density of fluid, kg/m³
E exchanger thermal effectiveness	N_C number of channels	W mass flow rate, kg/s	η dimensionless constant
f Fanning friction factor	Nu Nusselt number, $Nu = hDe/k$	x channel fluid flow tangential coordinate, m	θ dimensionless temperature of fluid
g gravitational acceleration, $g = 9.8$ m/s²	P number of passes	Y_f binary constant for channel-flow	ϕ constant for Feed contact respective location
G_C channel mass velocity, kg/m² s	Pr Prandtl number, $Pr = C_p\mu/k$	Y_h binary constant for hot fluid	Φ plate area enlargement factor

16.3 MATHEMATICAL MODEL FOR PHE

The aim of establishing this model is to analysis various effects of the parameters on the proficiency of heat exchanger.

16.3.1 CONFIGURATION CHARACTERIZATION

The parameters N_C, P^I, P^{II}, ϕ, Y_h, and Y_f are explained in Table 16.1:

N_C, the number of channels used in exchanger; P^I and P^{II}, the number of passes used in exchanger at sidewise 1 and 2; ϕ, the feed contact respective location (the feed contact of side 1 is arranged to channel 1 at $x = 0$); Y_h, the hot fluid position (it is a two-fold specification having a value 0 or 1, which allows the fluids toward exchanger sidewise. If $Y_h = 1$, hot fluid located at sidewise 1, and cold fluid located at sidewise 2); Y_f, the nature of flow in channels. (if $Y_f = 1$, in all channels the flow is diagonal. If $Y_f = 0$, in all channels the flow is vertical.)

TABLE 16.1   Correlation between Channels and Passes Numbers.

Premier relations		
$N_C = N_C^I + N_C^{II}$	$N_C^I = N_C \times P^I$	$N_C^{II} = N^{II} \times P^{II}$
If N_C is even		If N_C is odd
$N_C^I = \dfrac{N_C}{2}$ $\qquad N^I = \dfrac{N_C}{2P^I} N_C^I$	$N_C^I = \dfrac{N_C + 1}{2}$	$N^I = \dfrac{N_C + 1}{2P^I}$
$N_C^{II} = \dfrac{N_C}{2}$ $\qquad N^{II} = \dfrac{N_C}{2P^{II}} N_C^{II}$	$N_C^{II} = \dfrac{N_C - 1}{2}$	$N^{II} = \dfrac{N_C - 1}{2P^{II}}$

16.3.2   MODELING EQUATIONS FOR PHE

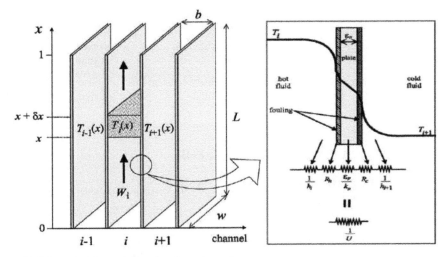

FIGURE 16.1   Upward direction flow channel.

By applying energy balance in Figure 16.1, we get:

$$\frac{dT_i}{dx} = \frac{s_i w \varnothing U_{i-1}}{W_i C_{pi}} \times \left(T_{i-1} - T_i\right) + \frac{s_i w \varnothing U_i}{W_i C_{pi}} \times \left(T_{i+1} - T_i\right) \tag{16.1}$$

By putting a parameter S_i, one can express the direction of flow of fluid in eq 16.1, that is, $S_i = +1$ for rising direction and $S_i = -1$ for falling direction in channel.

$$W_i = \frac{W^{side(i)}}{N^{side(i)}}, \qquad i = 1........N_c, \qquad side(i) = \{I, II\} \qquad (16.2)$$

U_i lies in the middle of channel i and $i + 1$ and is defined as in eq 16.3

$$\frac{1}{U_i} = \frac{1}{h_i} + \frac{1}{h_{i+1}} + \frac{\varepsilon_p}{k_p} + R_h + R_c, \qquad i = 1,.......(N_c - 1) \qquad (16.3)$$

The distance of way x and temperature of fluid in channel, $T_i (X)$ are transformed toward dimensionless parameter. So, we define two dimensionless parameters, that is, η and θ.

$$\eta(x) = X/L, \qquad 0 \le \theta \le 1 \qquad (16.4)$$

$$\frac{T_i - T_{c,in}}{T_{h,in} - T_{c,in}} \qquad 0 \le \theta \le 1 \qquad (16.5)$$

$$\frac{d\theta_i}{d\eta} = \frac{s_i(w\Phi L)U_{i-1}}{W_i C_{p_i}} \times \frac{(T_{i-1} - T_i)}{(T_{h,in} - T_{c,in})} + \frac{s_i(w\Phi L)U_i}{W_i C_{p_i}} \times \frac{(T_{i+1} - T_i)}{(T_{h,in} - T_{c,in})} \qquad (16.6)$$

Let $A = \Phi L$, the area enlargement factor

$$\frac{d\theta_i}{d\eta} = \frac{s_i A}{W_i C_{p_i}} \left[U_{i-1} \times (\theta_{i-1} - \theta_i) + U_i \times (\theta_{i+1} - \theta_i) \right] \qquad 1 < i < N_c \qquad (16.7)$$

Equation 16.7 is the general dimensionless equation for the i-th channel. So from this general equation we can write the equation for any channel as follows. If we assume the physical properties to be constant then the overall heat transfer coefficient can be assumed to be constant.

$$\therefore \frac{d\theta_1}{d\eta} = \frac{s_1 AU}{W_1 C_{p_1}} (\theta_2 - \theta_1) \qquad (16.8a)$$

$$\frac{d\theta_i}{d\eta} = \frac{s_i AU}{W_i C_{p_i}} (\theta_{i-1} - 2\theta_i + \theta_{i+1}) \qquad 1 < i < N_c \qquad (16.8b)$$

$$\frac{d\theta_{N_c}}{d\eta} = \frac{S_{N_c} AU}{W_{N_c} C_{p_i}} (\theta_{N_c - 1} - \theta_{N_c}) \qquad (16.8c)$$

If channel one is assigned as side I and channel two as side II, then defining α as given below:

$$\alpha^{I} = \frac{AUN^{I}}{W^{I}C^{I}_{p_{in}}} \quad ; \quad \alpha^{II} = \frac{AUN^{II}}{W^{II}C^{II}_{p_{in}}} \tag{16.9}$$

∴ By using above correlation the final equations are estimated:

$$\frac{d\theta_{1}}{d\eta} = s_{1}\alpha^{I}\left(\theta_{2} - \theta_{1}\right) \qquad \text{for first channel (I)} \tag{16.10a}$$

$$\frac{d\theta_{1}}{d\eta} = s_{1}\alpha^{side(t)}\left(\theta_{i-1} - 2\theta_{i} + \theta_{i+1}\right) \quad \begin{array}{l}\text{channel: } (1 < i < N_{c}), \\ \text{side (i)} = \{I, II\}(II)\end{array} \tag{16.10b}$$

$$\frac{d\theta_{N_{C}}}{d\eta} = s_{N_{C}}\alpha^{side(N_{C})}\left(\theta_{N_{C}-1} - \theta_{N_{C}}\right) \quad \begin{array}{l}\text{last channel, side } (N_{c}) \\ = \{I, II\}(II)\end{array} \tag{16.10c}$$

Consequently, eqs 16.8a–16.8c are compressed to a linear system of ODE eqs 16.10a–16.10c, which can be solved by Numerical Method (R-K 4th order method and finite difference method). The thermal boundary conditions for this PHE model (eq 16.10) are mentioned in Table 16.2.

TABLE 16.2 Boundary Condition for the Model PHE.

Boundary condition	Equation form
Fluid entrance	$\theta(\eta) = \theta_{\text{fluid, in}}, \qquad i \in$ first pass
Change of pass	$\theta_{i}(\eta) = \dfrac{1}{N}\displaystyle\sum_{j \in \text{previous pass}}^{N} \theta_{j}(\eta)$
Fluid outlet	$\theta_{\text{fluid,out}} = \dfrac{1}{N}\displaystyle\sum_{j \in \text{last pass}}^{N} \theta_{j}(\eta)$

Estimation of the parameters E and ΔP is essential for the analysis of the performance of gasketed PHE. The overall conservation of energy denotes that $E = E^{I} = E^{II}$

$$E^{I} = \frac{W^{I}C^{I}_{pm}\left|\theta_{in} - \theta_{out}\right|^{I}}{\min\left(W^{I}C^{I}_{pm}, W^{2}C^{2}_{pm}\right)} \quad ; \quad E^{II} = \frac{W^{II}C^{II}_{pm}\left|\theta_{in} - \theta_{out}\right|^{II}}{\min\left(W^{I}C^{I}_{pm}, W^{2}C^{2}_{pm}\right)} \tag{16.11}$$

The pressure drop of fluid at sides 1 and 2, ΔP^I and ΔP^{II} can be estimated by eq 16.12

$$\Delta P = \left(\frac{2f(1+D_P)PG_P^2}{\rho_m D_e}\right) + 1.4\left(P\frac{G_P^2}{2\rho_m}\right) + \rho_m g(L+D_P) \qquad (16.12)$$

$$G_C = \frac{W}{Nbw}; \quad G_P = \frac{4W}{\Pi D_P^2} \text{ (for sides 1 and 2)}; \quad D_e = \frac{4bw}{2(b+w\emptyset)} \approx \frac{2b}{\emptyset}$$

The interconnection for the estimation of h and f is given in eq 16.13. The general values for constants a_1 to a_6 are given in Table 16.3.

$$Nu = a_1 \text{Re}^{a_2}; f = a_4 + \frac{a_5}{\text{Re}^{a_6}} \qquad \text{for sides 1 and 2} \qquad (16.13)$$

TABLE 16.3 Input Conditions for Sugar Solution and Clean Fluid (SI unit).

Hot fluid (sucrose sol. 60° Brix)	Cold fluid (clean water)
$T_{in,hot} = 35°C$	$T_{in,cold} = 1°C$
$W_{hot} = 1.30$	$W_{cold} = 1.30$
$R_{f,hot} = 8.6e{-}05$	$R_{f,cold} = 1.7e{-}05$
$\rho_{hot} = 1286$	$\rho_{cold} = 1000$
$C_{phot} = 2803$	$C_{pcold} = 4206$
$K_{hot} = 0.407$	$k_{cold} = 0.584$
$a_{1,hot} = 0.400$	$a_{1,cold} = 0.3000$
$a_{2,hot} = 0.598$	$a_{2,cold} = 0.6633$
$a_{3,hot} = 0.3333$	$a_{3,cold} = 0.3333$
$a_{4,hot} = 0.0000$	$a_{4,cold} = 0.0000$
$a_{5,hot} = 18.299$	$a_{5,cold} = 1.4411$
$a_{6,hot} = 0.6521$	$a_{6,cold} = 0.206$

16.4 SOLUTION STRATEGY

The analytical modeling of a gasketed PHE, estimating E and ΔP of fluid, has been discussed before. However, it is not possible to obtain a model that is apparent of the configuration parameters (Table 16.4), conspicuously because of two-fold changeable S_i and the boundary condition given in Table 16.2.

TABLE 16.4 Instance Data for the Simulation.

Plates (stainless 360, chevron)	
L	0.75 m
W	0.235 m
B	0.0026 m
D_p	0.058 m
B	45°
Φ	1.17
ε_p	0.0007 m
K_p	17 W/m²°C
Configuration	
N_C	4
P^I	2
P^{II}	2
Φ	3
Y_h	1
Y_f	1

16.4.1 SIMULATION OF GENERAL PLATE HEAT EXCHANGER MODEL USING CONFIGURATION

This model of the gasketed PHE is developed by simulating it at the steady-state condition. The design steps of the PHE model are given below.

1. Observe the instance data for the PHE for hot and cold fluids and the configuration parameters.
2. The constant Y_h designates all fluid data at one and the other sides.
3. The number of channels for each pass at one and the other sides is estimated by N_C (see Table 16.1).

4. Estimation of ΔP for one and the other sides of the PHE is done using eq 16.12.
5. Coefficients α^I and α^{II} are estimated by eq 16.9, applying U from eq 16.3.
6. The values of two-fold changeable S_i are decided.
7. Equations are arranged, beginning with the differential equations toward the central channels (eq 16.10b).
8. The differential equations at the beginning and closing channels are also inserted in the system (eqs 16.10a and 16.10c, respectively).
9. Boundary conditions at the channels in one and the other sides are accomplished as shown in Table 16.2.
10. Equations of E at one and the other sides are estimated from eq 16.11.
11. The other equations are determined by numerical methods.
12. The simulated results like as ΔP and T_{outlet} at both sides (using parameter Y_h) and PHE thermal effectiveness are obtained.

In this study, the Runge–Kutta method of fourth order and finite difference method solved by MATLAB have been applied in eq 16.10b, using the boundary conditions mentioned in Table 16.2.

16.5   RESULTS AND DISCUSSION

The results are accomplished by simulating the model of the gasketed PHE with the input data being taken from Tables 16.3 and 16.4. A simulation program for the system of ODEs 45 based on for RK-4 numerical approach and finite difference method using five nodes has been developed in MATLAB. The approach so used in this study is applied to the PHE-containing process fluid of sugar solution and clean water. In this model, hot fluid using sugar solution flowing sideway (1) and cold fluid using clean water flowing sideway (2) were considered.

For the entire study, the obtained simulation results, that is, the pressure drop, thermal effectiveness, and outlet temperature, are shown in Table 16.5. The variation of channel temperature, plate temperature, fluid temperature, and the overall heat transfer coefficient with respect to plate length are presented in Figures 16.2–16.5. One can see a little bit deviation in the results of both methods. After comparison of the results, it can be concluded that Runge–Kutta method is slightly better than finite difference method.

TABLE 16.5 Results Obtained by PHE Model.

Variables	Runge–Kutta 4th order methods	Finite difference method	Deviation (%)
PHE thermal effectiveness (%)	75.61	74.97	0.8
Overall heat transfer coefficient, U (W/m² T°C)	1054	1049	0.47
Sucrose solution pressure drop (kPa)	76.9	76.9	–
Water pressure drop (kPa)	23.1	23.1	–
Hot fluid (Sucrose solution) outlet temperature (T,°C)	10.5	10.6	0.09
Cold fluid (clean water) outlet temperature (T,°C)	17.9	18.1	1
CPU time (s)	6	6.2	–

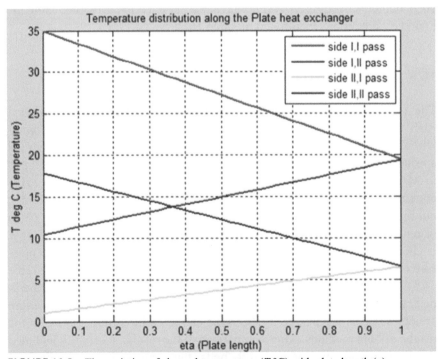

FIGURE 16.2 The variation of channel temperature (T,°C) with plate length (η).

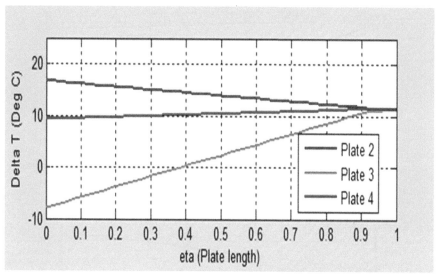

FIGURE 16.3   The variation of plate temperature (ΔT,°C) with plate length (η).

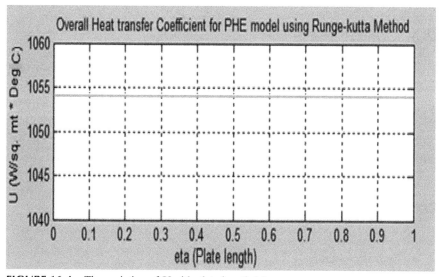

FIGURE 16.4   The variation of U with plate length (η).

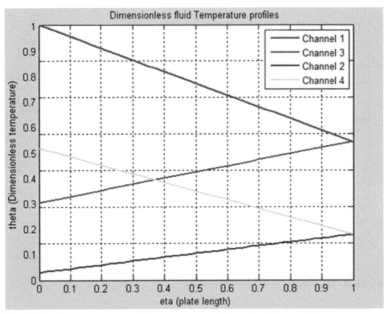

FIGURE 16.5   The variation of fluid temperature (θ) with plate length (η).

16.6   CONCLUSIONS

The results show small influence over the primary simulation result over the heat exchanger (thermal effectiveness, pressure drop, and outlet fluid temperatures) as far as U is concerned. One can see a little bit deviation in the results of both methods, as the Runge–Kutta method is slightly better than finite difference method. This program is a substantial tool for analyzing the effects of the arrangement in the exchanger accomplishment.

KEYWORDS

- **heat exchanger**
- **respective location**
- **dimensionless parameter**
- **configuration parameters**
- **difference method**

REFERENCES

1. Pignotti, A.; Tamborenea, P. I. Thermal Effectiveness of Multipass Plate Exchangers. *Int. J. Heat Mass Trans.* **1988,** *31,* 1983–1991.
2. Settari, A.; Venart, J. E. S. Approximate Method for the Solution to the Equations for Parallel and Mixed-Flow Multi-Channel Heat Exchangers. *Int. J. Heat Mass Trans.* **1972,** *15,* 819–829.
3. Ribeiro Jr., C. P.; Cano Andrade, M. H. An Algorithm for Steady-State Simulation of Plate Heat Exchangers. *J. Food Eng.* **2002,** *53,* 59–66.
4. Dardour, H.; Mazouz, S.; Bellagi, A. Numerical Analysis of Plate Heat Exchanger Performance in Co-Current Fluid Flow Configuration. *Int. J. Aerospace Mech. Eng.* **2009,** *3,* 141–145.
5. Qiao, H.; Aute, V. A New Model for Plate Heat Exchangers with Generalized Flow Configurations and Phase Change. *IJR.* **2013,** *36,* 622–632.
6. Pinto, J. M.; Gut, J. A. W. Modeling of Plate Heat Exchanger with Generalized Configuration. *Int. J. Heat Mass Trans.* **2003,** *46,* 2571–2585.
7. Pinto, J. M.; Gut, J. A. W. Thermal Model Validation of Plate Heat Exchangers with Generalized Configurations. *Chem. Eng. Sci.* **2004,** *59,* 4591–4600.
8. Georgiadis, M. C.; Macchietto, S. Dynamic Modeling and Simulation of Plate Heat Exchangers under Milk Fouling. *Chem. Eng. Sci.* **2000,** *55,* 1605–1619.
9. Bassiouny, M. K.; Martin, H. Flow Distribution and Pressure Drop in Plate Heat Exchangers-I. *Chem. Eng. Sci.* **1984,** *39,* 693–700.
10. Kandlikar, S. G.; Shah, R. K. Multipass Plate Heat Exchangers-Effectiveness-NTU Results and Guidelines for Selecting Pass Arrangements. *J. Heat Trans.* **1989,** *111,* 300–313.
11. Zaleski, T. A General Mathematical-Model of Parallel-Fow, Multichannel Heat-Exchangers and Analysis of its Properties. *Chem. Eng. Sci.* **1983,** *39,* 1251–1260.
12. Zaleski, T.; Klepacka, K. Approximate Methods of Solving Equations for Plate Heat-Exchangers. *Int. J. Heat Mass Trans.* **1992,** *35,* 1125–1130.
13. Shah, R. K.; Focke, W.W. Plate Heat Exchangers and their Design Theory. In *Heat Transfer Equipment Design;* Shah, R. K., Subbarao, D. C., Mashelekar, R. M., Eds.; Hemisphere: Washington, D. C., 1988; pp 227–254.

CHAPTER 17

OPTIMIZATION OF ADSORPTION CAPACITY OF PREPARED ACTIVATED CARBON USING RESPONSE SURFACE METHODOLOGY

V. N. GANVIR[1,*], M. L. MESHRAM[2], and R. R. PATIL[1]

[1]Department of Petrochemical Technology, Laxminarayan Institute of Technology, Nagpur, Maharashtra, India

[2]Department of Civil Engineering, Laxminarayan Institute of Technology, Nagpur, Maharashtra, India

*Corresponding author. E-mail: vnganvirlit@gmail.com

CONTENTS

ABSTRACT

Activated carbon was prepared from karanja seed shell. Chemical activation of precursor was done using sulfuric acid as an activating agent. Adsorption capacity of prepared activated carbon was examined by adsorption of methylene blue. Optimization of adsorption capacity of the prepared activated carbon was studied using response surface methodology (RSM). In RSM, central composite design (CCD) was used to find the quadratic response surface equation. Analysis of variance (ANOVA) of the quadratic model suggested that the predicted values were in good agreement with the experimental data. A model equation for predicting the response was formulated which can be successfully adopted in industry to maximize the response of the process or operation. The experimental results indicated that the optimal conditions for the maximum adsorption capacity were found to be 125 mg/L and 210.229 min for methylene blue concentration and contact time, respectively. Under these conditions, the maximum adsorption capacity was found to be 31.2754 mg/g.

17.1 INTRODUCTION

Adsorption is a well-known separation operation/technique widely used to remove certain classes of pollutants from wastewater and to remove heavy metals from industrial waste streams.[1] Activated alumina, silica gel, activated carbon, molecular sieve (carbon and zeolite), and polymeric materials are commonly used as adsorbents. Activated carbon is an excellent adsorbent. To remove heavy metals from ground water, impurities from effluents, pollutants from waste streams of industry, the adsorbent must have high surface area, good pore size distribution, as well as suitable mechanical properties.[2-4]

In recent years, crop or agricultural wastes are used as the source to prepare activated carbon. Some of them are rice husk,[5] sawdust,[6] maize wastes,[7] neem bark,[8] pine wood,[9] coconut shell,[10] peanut hulls,[11] orange peel,[12] banana peel,[13] corn cob,[14] mango seed,[15] and cassava peel.[16] Industrial wastes such as tea industry waste,[17] fly ash, blast furnace slag, and sludge[18] are also used to prepare activated carbon.

Demand of activated carbon as an adsorbent is low due to its high processing cost. Generally, low-cost activated carbon are those which require less processing, are easily available, and are abundant in nature.[19]

Pongamia pinnata is one such species whose oil is widely used in biodiesel industry/vegetable oil refinery. Since it is easily available in India it can be used as a source to prepare low-cost adsorbent.[20] Recently much research has been done by scientists/researchers to prepare low-cost adsorbent from renewable sources.

Mostly dyes are used in textile industries. Wastewater and effluents from textile industries contain dyes which are carcinogenic and toxic in nature, and their degradation takes a lot of time. Therefore, the removal of such dyes from process effluent becomes environmentally important. Removal of dye molecules from waste streams of dye industries will be a cost effective process if low-cost activated carbon is utilized as an adsorbent.[21,22]

Response surface methodology (RSM) is a mathematical modeling based on statistical techniques and is a useful tool to optimize chemical process/operation.[23–25]

In this study, activated carbon is prepared by chemical activation of *P. pinnata* seed shell (karanja) and its application is studied by adsorbing methylene blue on it. The RSM is used to optimize adsorption capacity of prepared activated carbon. These were the objectives of this study.

17.2   MATERIALS AND METHODS

17.2.1   CHEMICAL ACTIVATION OF PONGAMIA PINNATA SEED SHELL (KARANJA SEED SHELL)

P. pinnata seed shell was used as the feed to prepare activated carbon. Karanja beans were collected from the campus of Laxminarayan Institute of Technology, Nagpur, India. Karanja seed shells were dried in sunlight for 15–20 days to remove moisture. Chemical activation (concentrated sulfuric acid) was done in the impregnation ratio of 1:5 followed by carbonization (temperature 850°C and time 10 min).[26]

17.2.2   PREPARATION OF METHYLENE BLUE SOLUTION

Methylene blue (1 g) was dissolved in 1000 mL of distilled water for the preparation of stock solution.[27] From stock solution, solutions of different concentrations 50,100, and 150 mg/L were prepared.

17.2.3 STUDY OF ADSORPTION OF METHYLENE BLUE OF PREPARED ACTIVATED CARBON

Activated carbon 0.2 g were mixed with 50 mL of methylene blue solution with an initial concentration ranging between 50 and 150 mg/L. The difference in concentration was measured with UV–Vis spectrophotometer.[28–29] The removal efficiency was calculated using the following equation:[30,31]

$$\text{Removal efficiency (\%)} = \frac{C_i - C_o}{C_o} \times 100,$$

where C_i is the initial concentration of methylene blue and C_o is its final concentration.

The adsorption capacity was calculated using this equation:

$$\text{Adsorption capacity (mg/g)} = C_i - C_o \times \frac{V}{M}$$

where M is the weight of adsorbent and V is the volume of methylene blue solution.

17.2.4 OPTIMIZATION OF ADSORPTION CAPACITY OF ACTIVATED CARBON (PREPARED) USING RSM

For optimization of adsorption capacity of adsorbent, central composite design (CCD) was used. The parameters for adsorption capacity study were methylene blue concentration and contact time.[32] Generally the CCD is based on following equation: $N = a + b + c$, where $N =$ total number of runs, $a = 2^n$ factorial runs, $b = 2n$ axial runs, and $c = n_c$ centers runs (six replicates). The variables studied are

1. X_1: methylene blue concentration (mg/L),
2. X_2: contact time (min).

Thus total number of runs (N) required is 14:

$$N = 2^n + 2n + n_c = 2^2 + 2 \times 2 + 6 = 14$$

The center points are used to determine the experimental error and the reproducibility of the data.[33,34] Table 17.1 shows the effect of methylene blue concentration and contact time (input parameters) on adsorption capacity (response) of activated carbon.

TABLE 17.1 Effect of Methylene Blue Concentration and Contact Time on Adsorption Capacity.

Sr. No.	Concentration of methylene blue (mg/L)	Contact time (min)	Experimental adsorption capacity (mg/g)
1	50	15	12.28
2	50	30	12.39
3	50	45	12.42
4	50	60	12.46
5	50	90	12.49
6	100	60	24.27
7	100	90	24.74
8	100	120	24.86
9	100	180	24.88
10	100	240	25.03
11	150	180	37.27
12	150	240	37.57
13	150	300	37.75
14	150	360	37.82

The linear equation in model is given as

$$Y = A + (B \times X_1) + (C \times X_2) \tag{17.1}$$

Similarly, the cross equation in model is

$$Y = A + (B \times X_1) + (C \times X_2) + (D \times X_1 \times X_2) \tag{17.2}$$

And, the square equation is represented in model as

$$Y = A + (B \times X_1) + (C \times X_2) + (D \times X_1 \times X_2) + (E \times X_1 \times X_1) + (F \times X_2 \times X_2) \tag{17.3}$$

where X_1 and X_2 are independent variables, whereas A, B, C, D, E, and F are the unknown values in the equations. The unknown values for the above equations are calculated using software (LABFIT, trial version online freely available) or MATLAB.

17.3 RESULTS AND DISCUSSION

17.3.1 DATA ANALYSIS

The data are analyzed using models: namely linear, cross, and square. The values of regression correlation for linear, cross, and square values for output Y are calculated with the help of the following equations.

Output Y is calculated by putting the values of A, B, and C in eq 17.1; therefore the equation becomes:

$$Y = 0.625816169900 + 0.2526949244278 \times X_1 + 0.006322664172946 \times X_2 ... R^2 = 0.9837.$$

Output Y is calculated by putting values of A, B, C, and D in eq 17.2; therefore the equation becomes

$$Y = -0.3069586885991 + 0.2491368128817 \times X_1 - 0.001079803404641 \times 0.00008552137479389 \times X_1 \times X_2 ... R^2 = 0.9952.$$

Output Y is calculated by putting the values of A, B, C, D, E, and F in eq 17.3; therefore the equation becomes

$$Y = 0.5918904847309 + 0.2227311060081 \times X_1 + 0.008615121835668 \times X_2 + 0.00009889724636608 \times X_1 \times X_2 + 0.0001298788411074 \times X_1 \times X_1 - 0.00009967074992948 \times X_2 \times X_2 ... R^2 = 0.9066,$$

where X_1 and X_2 are independent variables, that is, concentration of methylene blue (X_1), time (X_2). Whereas A, B, C, D, E, and F are the unknown values in the equations. For model equation, value of regression coefficient (R^2) near to 1 was selected. Here the value of regression coefficient (R^2) for cross equation (eq 17.2) is 0.9952 which is near to 1 than square (eq 17.3) and linear (eq 17.1) type equations, and the predicted values (Table 17.2) of response for cross type equation were developed in MS Excel sheet.

TABLE 17.2   Experimental Design for the Adsorption of Methylene Blue Solution onto Adsorbent and Comparison of Actual and Predicted Responses.

Sr. No.	Concentration of methylene blue (mg/L)	Contact time (min)	Experimental adsorption capacity (mg/g)	Predicted adsorption capacity (mg/g)
1	50	15	12.28	12.3153
2	50	30	12.39	12.3689
3	50	45	12.42	12.4134
4	50	60	12.46	12.4490
5	50	90	12.49	12.4932
6	100	60	24.27	24.4100
7	100	90	24.74	24.6040
8	100	120	24.86	24.7620
9	100	180	24.88	24.9699
10	100	240	25.03	25.0337
11	150	180	37.27	37.2337
12	150	240	37.57	37.5973
13	150	300	37.75	37.8167
14	150	360	37.82	37.8921

TABLE 17.3   ANOVA for the Fitted Model.

Source	Sum of squares	Df	Mean square	F value	P value prob > F	Significant
Model	1211.82	3	403.94	120536.99	<0.0001	
X_1	325.74	1	325.74	97202.08	<0.0001	
X_2	0.36	1	0.36	106.98	<0.0001	
X_1X_2	0.13	1	0.13	39.63	0.0001	
Residual	0.030	9	0.003351			
Corr. total	1211.85	12				

Notes: X_1, methylene blue concentration; X_2, contact time; adjusted $R^2 = 1$; predicted $R^2 = 0.9999$; adequate precision = 794.301.

The obtained data were analyzed (in Design Expert 9 Stat ease trial version) using analysis of variance (ANOVA) (Table 17.3). The model F

value of 120536.99 implies the model is significant. Here X_1, X_2, and X_1X_2 are significant model terms. The "Predicted R-Squared" of 0.9999 is in reasonable agreement with the "Adjusted R-Squared" of 1.0000; that is a difference of <0.2. The ratio of 794.301 indicates an adequate signal.

ANOVA of the quadratic model suggested that the predicted values were in good agreement with experimental data.[35] A model equation for predicting the response was formulated which can be successfully adopted in industry to maximize the response of the process or operation (Fig. 17.1).

Model equation is as

Adsorption capacity = $24.84 + 12.27 \times X_1 + 0.49 \times X_2 + 0.27 \times X_1 \times X_2$.

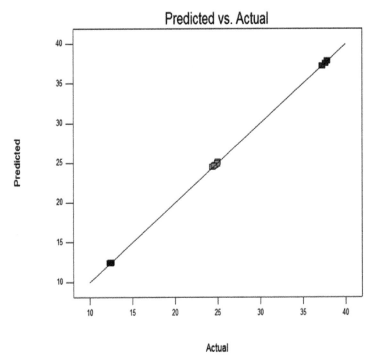

FIGURE 17.1   Predicted value vs. actual value.

Figs. 17.2 and 17.3 represent the combined effect of methylene blue concentration and contact time on the adsorption capacity of activated carbon. From the plot, the adsorption capacity was found to increase with of the rise of methylene blue concentration and contact time.

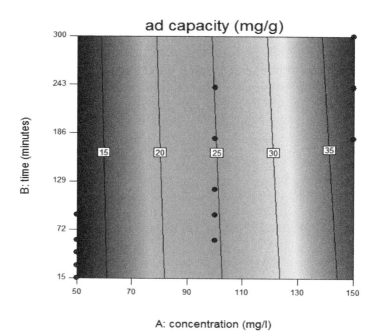

FIGURE 17.2 Contour plot of the combined effects of contact time and methylene blue concentration on adsorption capacity.

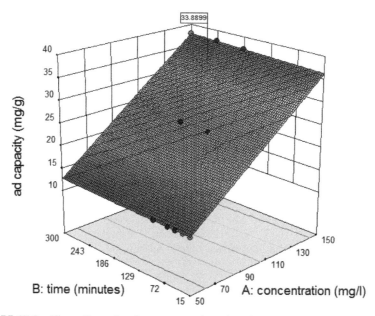

FIGURE 17.3 Three-dimensional response surface plot of adsorption capacity.

17.4 CONCLUSIONS

Karanja seed shell (*P. pinnata*) is found to be a good source to prepare activated carbon and its optimum condition for carbonization temperature and time were 850°C and 10 min, respectively. For the removal of methylene blue it could be used as an effective adsorbent. The experimental results indicated the adsorption capacity of activated carbon for concentration of methylene blue of 150 mg/L and contact time of 240 min was found to be 31.57 mg/g and using RSM, adsorption capacity for concentration of methylene blue 25 mg/L and contact time 210.229 min was found to be 31.2754 mg/g. The obtained model equation is,

$$\text{Adsorption capacity} = 24.84 + 12.27 \times X_1 + 0.49 \times X_2 + 0.27 \times X_1 \times X_2.$$

KEYWORDS

- adsorption
- dyes
- central composite design
- karanja
- methylene blue

REFERENCES

1. Ahmaruzzaman, M. Adsorption of Phenolic Compounds on Low-Cost Adsorbents: A Review. *Adv. Colloid Interf. Sci.* **2008,** *143,* 48–67.
2. Liu, Q.; Zheng, T.; Wang, P.; Guo, L. Preparation and Characterization of Activated-carbon from Bamboo by Microwave-Induced Phosphoric Acid Activation. *Ind. Crop. Prod.* **2010,** *31,* 233–238.
3. Zhong, Z.; Yang, Q.; Li, X.; Luo, K.; Liu, Y.; Zeng, G. M. Preparation of Peanut Hull-Based Activated Carbon by Microwave-Induced Phosphoric Acid Activation and its Application in Remazol Brilliant Blue R Adsorption. *Ind. Crops Prod.* **2012,** *37,* 178– 185.
4. Zeng, C.; Lin, Q.; Fang, C.; Xu, D.; Ma, Z. Preparation and Characterization of High Surface Area Activated Carbons from Co-Pyrolysis Product of Coal-Tar Pitch and Rosin. *J. Ana. App. Pyrolysis.* **2013,** *104,* 372–377.

5. Ajmal, M.; Rao, R. A. K.; Anwar, S.; Ahmad, J.; Ahmad, R. Adsorption Studies on Rice Husk: Removal and Recovery of Cd (II) from Wastewater. *Bioresour. Tech.* **2003**, *86,* 147–149.

6. Kadirvelu, K.; Kavipriya, M.; Karthika, C.; Radhika, M.; Vennilamani, N.; Pattabhi, S. Utilization of Various Agricultural Wastes for Activated Carbon Preparation and Application for the Removal of Dyes and Metal Ions from Aqueous Solution. *Bioresour. Tech.* **2003**, *87,* 129–132.

7. Jimenez D.; Elizalde-Gonzalez, M. P.; Pelaez-Cid, A. A. Adsorption Interaction between Natural Adsorbents and Textile Dyes in Aqueous Solution. *Colloids Surf. A.* **2005**, *254,* 107–114.

8. Ayub, S.; Ali, S. I.; Khan, N. A. Efficiency Evaluation of Neem (Azadirachtaindica) Bark in Treatment of Industrial Wastewater. *Environ. Poll. Control J.* **2001**, *4* (4), 34–38.

9. Tseng, R. L.; Wu, F. C.; Juang, R. S. Liquid-Phase Adsorption of Dyes and Phenols Using Pinewood-Based Activated Carbons. *Carbon.* **2003**, *41,* 487–495.

10. Namasivayam, C.; Dinesh-Kumar, M.; Selvi, K.; Ashruffunissa-Begur, R.; Vanathi, T.; Yamuna, R. T. Waste Coir Pith – a Potential Biomass for the Treatment of Dyeing Wastewaters. *Biomass. Bioenergy.* **2001**, *21,* 477–483.

11. Gong, R.; Ding, Y.; Li, M.; Yang, C.; Lui, H.; Sun, Y. Utilization of Powdered Peanut Hull as Biosorbent for Removal of Anionic Dyes from Aqueous Solution. *Dyes Pigm.* **2005**, *64,* 187–192.

12. Sivaraj, R.; Namasivayam, C.; Kadirvelu, K. Orange Peel as an Adsorbent in the Removal of Acid Violet 17 (Acid Dye) from Aqueous Solutions. *Waste Manag.* **2001**, *21,* 105–110.

13. Annadurai, G.; Juang, R. S.; Lee, D. J. Use of Cellulose-Based Wastes for Adsorption of Dyes from Aqueous Solutions. *J. Hazard. Mater.* **2002**, *B92,* 263–274.

14. Wu, F. C.; Tseng, R. L.; Juang, R. S. Adsorption of Dyes and Phenols from Water on the Activated Carbons Prepared from Corncob Wastes. *Environ. Technol.* **2001**, *22,* 205–213.

15. Davila-Jimenez, M. M.; Elizalde-Gonzalez, M. P.; Hernandez-Montoya, V. Performance of Mango Seed Adsorbents in the Adsorption of Anthraquinone and Azo Acid Dyes in Single and Binary Aqueous Solutions. *Biores. Tech.* **2009**, *100,* 6199–6206.

16. Sivaraj R.; Sivakumar, S.; Senthilkumar, P.; Subburam, V. Carbon from Cassava Peel, an Agricultural Waste, as an Adsorbent in the Removal of Dyes and Metal Ions from Aqueous Solution. *Bioresour. Tech.* **2001**, *80,* 233–235.

17. Ali, G.; Celal, D. H.; Basri, S.; Soylak, M.; Imamoglud, M.; Onal, Y. Physicochemical Characteristics of a Novel Activated Carbon Produced from Tea Industry Waste. *J. Ana. App. Pyrolysis.* **2013**, *104,* 249–259.

18. Ahmaruzzaman, M. Industrial Wastes as Low-Cost Potential Adsorbents for the Treatment of Wastewater Laden with Heavy Metals. *Adv. Colloid and Interface Sci.* **2011**, *166,* 36–59.

19. Hegazi, H. A. Removal of Heavy Metals from Wastewater Using Agricultural and Industrial Wastes as Adsorbents. *HBRC. J.* **2013**, *9,* 276–282.

20. Ganvir, V. N.; Meshram, M. L.; Dwivedi, A. P. In *Preparation of Adsorbent from Karanja Oil Seed Cake and its Characterization,* International Conference on Emerging Frontiers in Technology for Rural Area (EFITRA) Proceedings, International Journal of Computer Applications (IJCA): New York, Vol. 3, 2012; pp 18–19.

21. Hameed, B. H.; Ahmad, A. L.; Latiff, K. N. A. Adsorption of Basic Dye (Methylene Blue) onto Activated Carbon Prepared from Rattan Sawdust. *Dyes Pig.* **2007**, *75,* 143–149.

22. Sadeghi-Kiakhani, M.; Arami, M.; Gharanjig, K. Preparation of Chitosan-Ethyl Acrylate as a Biopolymer Adsorbent for Basic Dyes Removal from Colored Solutions. *J. Env. Chem. Eng.* **2013**, *1*, 406–415.

23. Cao, J.; Wub, Y.; Jin, Y.; Yilihan, P.; Huang, W. Response Surface Methodology Approach for Optimization of the Removal of Chromium(VI) by NH2-MCM-41. *J. Taiwan Inst. Chem. Eng.* **2014**, *45*, 860–868.

24. Bezerra, M. A.; Santelli, R. E.; Oliveira, E. P.; Villar, L. S.; Escaleira, L. A. Response Surface Methodology (RSM) as a Tool for Optimization in Analytical Chemistry. *Talanta.* **2008**, *76*, 965–977.

25. El-Gendy, N. S.; El-Gharabawy Abd Allah, S. A.; Salem, S. A.; Ashour, F. H.; Response Surface Optimization of an Alkaline Transesterification of Waste Cooking Oil. *Int. J. Chem. Tech. Res.* **2015**, *8*, 385–398.

26. Jain, M.; Garg, V. K.; Kadirvelu, K. Adsorption of Hexavalent Chromium from Aqueous Medium onto Carbonaceous Adsorbents Prepared from Waste Biomass. *J. Env. Mgt.* **2010**, *91*, 949–957.

27. Subramaniam, R.; Ponnusamy, S. K Novel Adsorbent from Agricultural Waste (Cashew NUT Shell) for Methylene Blue Dye Removal: Optimization by Response Surface Methodology. *Water Resour. Ind.* **2015**, *11*, 64–70.

28. Baccar, R.; Blánquez, P.; Bouzid, J.; Feki, M.; Attiya, H.; Sarrà, M. Modeling of Adsorption Isotherms and Kinetics of a Tannery Dye onto an Activated Carbon Prepared from an Agricultural By-Product. *Fuel Process. Tech.* **2013**, *106*, 408–415.

29. Crini, G. Non-Conventional Low-Cost Adsorbents for Dye Removal: A Review. *Biores. Tech.* **2006**, *97*, 1061–1085.

30. Gao, J.; Qin, Y.; Zhou, T.; Cao, D.; Xu, P.; Hochstetter, D.; Wang, Y. Adsorption of Methylene Blue onto Activated Carbon Produced from Tea (Camellia sinensis L.) Seed Shells: Kinetics, Equilibrium, and Thermodynamics Studies. *J. Zhejiang Univ. Sci. B. Biomed. Biotech.* **2013**, *7*, 650–658.

31. Crisafully, R.; Aparecida, M.; Milhome, L.; Cavalcante, R. M.; Silveira, E. R.; Keukeleire, D. D.; Nascimento, R. F. Removal of Some Polycyclic Aromatic Hydrocarbons from Petrochemical Wastewater using Low-Cost Adsorbents of Natural Origin. *Biores. Tech.* **2008**, *99*, 4515–4519.

32. Sharma, P. B.; Kumar, N. *New Frontiers in Biofuels;* Scitech Publications *(India)* Pvt. Ltd.: Chennai, India, 2009.

33. Patil, P. D.; Gude, V. G.; Mannarswamy, A.; Deng, S.; Cooke, P.; Munson-McGee, S.; Rhodes, I.; Lammers, P.; Nagamany, N. Optimization of Direct Conversion of Wet Algae to Biodiesel Under Supercritical Methanol Conditions. *Biores. Tech.* **2011**, *102*, 118–122.

34. Silva, G. F.; Camargo, F. L.; Ferreira, A. L. O. Application of Response Surface Methodology for Optimization of Biodiesel Production by Trans Esterification of Soyabean Oil with Ethanol. *Fuel Proc. Tech.* **2010**, *92*, 407–413.

35. Danish, M.; Hashim, R.; Ibrahim, M. N.; Othman, S. Optimized Preparation for Large Surface Area Activated Carbon from Date (Phoenix dactylifera L.) Stone Biomass. *Biomass Bioenergy.* **2014**, *61*, 167–178.

CHAPTER 18

MODELING OF ETHANOL/WATER SEPARATION BY PERVAPORATION MEMBRANE PROCESS

N. G. KANSE[1,*], R. P. BIRMOD[2], and S. D. DAWANDE[2]

[1]Department of Chemical Engineering, FAMT, Ratnagiri, Maharashtra, India

[2]Department of Chemical Engineering, Laxminarayan Institute of Technology, Nagpur, Maharashtra, India

*Corresponding author. E-mail: nitin_475@yahoo.co.in

CONTENTS

ABSTRACT

The separation of ethanol–water mixture was carried out by using Pervaporation membrane separation process over a full range of compositions at temperatures varying from 45 to 75°C, using polyvinyl alcohol (PVA) dense membrane. In this study a mathematical model has been developed and it is able to describe the behavior of Pervaporation process. The basic model equations are developed for one pass Pervaporation system and complete mixing model. The given model is based on extension of Flory–Huggins thermodynamics and is applied to the Pervaporation membrane separation of full range of ethanol/water (azeotropic mixture) mixture through PVA membrane. The experimental data were compared with model data. The presented data can be extended for study of mathematical modeling of Pervaporation for other binary systems.

18.1 INTRODUCTION

In the recent past, there has been significant effort in exploring Pervaporation technology in separation applications. Many researchers have reported mathematical model, simulation, and experimental results of Pervaporation process with different types of membranes. In recent past the separation of a variety of binary mixtures such as organic mixtures, aqueous mixtures, azeotropic mixture, etc., has been studied. Nowadays Pervaporation has an increasing industrial relevance among separation methods. Pervaporation is a membrane separation technology using a non-porous dense homogeneous polymeric film as selective separation barrier. Membrane separation techniques have considerable applications in environmental protection, energy, medication, and food.

Pervaporation membrane technology is a relatively new process that has elements in common with reverse osmosis and gas separation. In Pervaporation process, a liquid mixture is fed to one side of a membrane, and the permeate is removed from the other side in the form of vapor. The permeate side driving force is low vapor pressure generated by cooling and condensing the permeate vapor. The importance of Pervaporation membrane process is that the separation obtained is proportional to the rate of permeation of the components of the liquid mixture through the selective membrane. Therefore, Pervaporation offers the possibility of separating azeotropic mixtures that are difficult to separate by other separation process such as distillation. A schematic of Pervaporation process is shown in Figure 18.1.

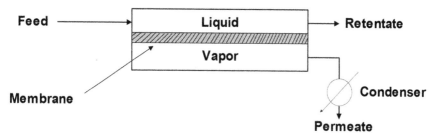

FIGURE 18.1 Schematic of pervaporation process.

In particular, Pervaporation membrane separation technology is developed more widely in industry to produce anhydrous ethanol because of its advantages within the separation process, such as requiring simple design, high single-stage separation efficiency, no addition of a third component, low energy consumption, and pollution free.[1-5] There are different model equations[6-9] for single component permeation which show more or less good agreement between theory and experiment.

The main objective of this study is to develop a mathematical model for the Pervaporation separation of ethanol–water mixture through a polymer membrane, taking into account the coupling of fluxes. Fels and Huang[9,10] developed a model for the Pervaporation membrane process by Fujita's free volume theory which is accepted for determining diffusion coefficients from unsteady-state desorption experiments. For explaining the interaction between permeate and membrane material they used simplified Flory–Huggins thermodynamic theories.

Mulder and Smolders[11,12] proposed a modified solution-diffusion model which describes the transport and concentration profiles of ethanol and water in dense membranes in their model. The exponential relationship between concentration and diffusion coefficient was used, and also included the dissolving ability of the mixture.

18.2 MATHEMATICAL MODELING OF PERVAPORATION PROCESS

The model follows the principle of the solution-diffusion model. Following assumptions have been made to simplify the model of Pervaporation process.

1. Equilibrium sorption exists during Pervaporation.
2. Swelling effect is negligible on the polymer morphology.
3. Temperature is constant throughout the system.

4. The interface of the membrane is in equilibrium with the upstream and downstream phase.
5. At permeate side of the membrane, concentration of the penetrants is zero.

In the Pervaporation of a pure liquid through a nonporous homogeneous polymer membrane, the transport occurs by a solution-diffusion-desorption mechanism. Fick's first law with a concentration dependent diffusion coefficient is used to describe steady-state permeation of a single component through a membrane.

$$J_k = -D_k \frac{dC_k}{dz},$$ (18.1)

where J_k is flux (g/cm²·s) and dC_k/dz is the concentration gradient of component k and D_k is diffusion coefficient (cm²/s). The minus sign for direction of diffusion is down the concentration gradient. Diffusion is a slow process. In practical diffusion-controlled separation processes, useful fluxes across the membrane are achieved by casting the membranes very thin as well as creating high concentration gradients in the membrane. The separation factor (α) and Pervaporation separation index (PSI) are evaluated as follows

$$\alpha = \frac{y_{water} / y_{ethanol}}{x_{water} / x_{ethanol}}$$ (18.2)

$$PSI = total\ flux \times (\alpha - 1)$$ (18.3)

where y_{water} and $y_{ethanol}$ are the weight fractions of water and ethanol in the permeate and x_{water} and $x_{ethanol}$ are the weight fractions of water and ethanol in the feed, respectively.

The gradient in the concentration of the liquid within the membrane is introduced by the chemical potential difference, which is maintained across the system.

$$\mu_1 - \mu_2 = RT \ln \frac{P_1}{P_2}$$ (18.4)

P_1 and P_2 are the equilibrium vapor pressure of the liquid and partial pressure of vapor, respectively.

When polymer is in contact with a solution, the diffusion phenomenon becomes highly non-ideal. For such cases the approach of modified Fickian diffusion with the gradient of chemical potential as the driving force holds good.[11]

$$J_k = -cB \frac{d\mu^m}{dz} \tag{18.5}$$

where

$$d\mu^m = RT d\ln a^m \tag{18.6}$$

Put the value of eq 18.6 in eq 18.5 we get

$$J_k = -cBRT \frac{d\ln a^m}{dz} \tag{18.7}$$

where $\bar{D} = BRT$ is known as thermodynamic diffusion coefficient of a species in a frame of reference stationary with respect to the polymer. Moreover in the Pervaporation membrane the solute diffusivity depends on the concentration in the polymer. The higher will be the extent of swelling, the more will be the concentration of the sorbed components and higher will be the diffusion coefficients. Equation 18.7 can be written as,

$$J_k = -\bar{D}(c_1, c_2) \frac{d\ln a^m}{d\ln c} \frac{dc}{dz} \tag{18.8}$$

For the components in a binary mixture (ethanol–water mixture)

$$J_w = -\bar{D}(c_w, c_e) \frac{d\ln a_w^m}{d\ln c_w} \frac{dc_w}{dz} \tag{18.9}$$

$$J_e = -\bar{D}(c_w, c_e) \frac{d\ln a_e^m}{d\ln c_e} \frac{dc_e}{dz} \tag{18.10}$$

The integration of the above two equations over the membrane thickness using the equilibrium concentration at the membrane surface at upstream and downstream sides from eq 18.9 and 18.10 gives the fluxes of the two components as well as the separation factor. The developed equations are highly nonlinear because of the concentration dependence of the diffusivities as well as the nonlinear dependence of the component activities on concentrations.

18.2.1 ONE PASS PERVAPORATION SYSTEM

Figure 18.2 shows a schematic of one pass Pervaporation system for a binary separation (ethanol/water).

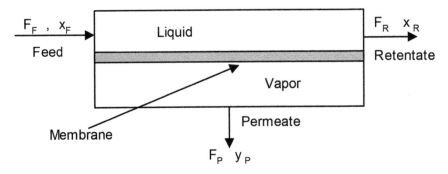

FIGURE 18.2   One pass pervaporation system.

The overall material balance
  Input − output = 0

$$F_F - F_R - F_P = 0 \tag{18.11}$$

where F_F is total feed flow rate, F_R is the outlet reject flow rate, and F_P is outlet permeate flow rate while x_F, x_r, and y_p are mole fractions of the more permeable component.
  The cut or fraction of feed permeated θ is given as,

$$\theta = \frac{F_P}{F_F} \tag{18.12}$$

Making overall material balance on component we get,

$$F_F x_F - F_R x_R - F_P y_P = 0 \tag{18.13}$$

The value of cut or fraction of feed permeated θ is often controlled by the energy balance.

The energy balance equation for one pass Pervaporation system is,

$$F_F c_{PF}\left[T_F - T_{ref}\right] = F_R c_{PR}\left[T_R - T_{ref}\right] + F_P[c_{PP}\left(T_P - T_{ref}\right) + \lambda] = 0 \quad (18.14)$$

where λ is the latent heat of vaporization.

Divide eq 18.13 by F_F we get

$$x_F - \frac{F_R x_R}{F_F} - \frac{F_P y_P}{F_F} = 0 \tag{18.15}$$

Solve eq 18.15 for the outlet rejected composition we get,

$$x_F - \frac{F_R x_R}{F_F} - \theta y_P = 0 \tag{18.16}$$

$$x_R = \frac{x_F - \theta y_P}{F_R / F_F} \tag{18.17}$$

But $F_R/F_F = 1 - \theta$, we get

$$x_R = \frac{(x_F - \theta y_P)}{1 - \theta} \tag{18.18}$$

$$y_P = \frac{x_F - x_R(1 - \theta)}{\theta} = 0 \tag{18.19}$$

The rate of diffusion of permeation of species A is given by an equation,

$$\frac{V_A}{A_m} = \frac{V_P y_P}{A_m} = \left(\frac{P'_A}{t}\right)(P_h x_R - P_L y_P) \tag{18.20}$$

Substituting θ in above equation we get

$$A_m = \frac{\theta F_F y_P}{(P'_A / t)(P_h x_R - P_L y_P)} \tag{18.21}$$

From eq 18.21 we can calculate membrane area A_m.

18.2.2 THERMODYNAMIC CONSIDERATIONS

The transport in Pervaporation can be comparable to that of reverse osmosis and gas permeation based on the solution-diffusion model under isothermal condition. According to this approach chemical potential gradient across the membrane is the driving force for mass transfer. Consider transport of species water (w) and ethanol (e) through membrane. The profile of pressure gradient is represented by Figure 18.3. Pressure drop from the upstream (feed side) pressure to the downstream of the membrane discontinuously.

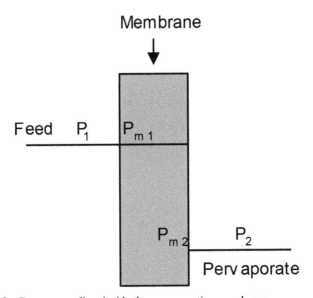

FIGURE 18.3   Pressure gradient inside the pervaporation membrane.

Chemical potential gradient for the species water (w) and ethanol (e) can be written as,

$$\Delta \mu w = RT \Delta \ln a_{wm} + v_w \Delta P \tag{18.22}$$

$$\Delta \mu e = RT \Delta \ln a_{em} + v_e \Delta P \tag{18.23}$$

where v_w and v_e are the molar volume (m^3/mol) of water and ethanol. Assumed that pressure P remains constant, the same as that of feed solution throughout the membrane cross-section flux equation of water and ethanol are,

$$J_{water} = -D_{wm}\frac{\Delta Cw_m}{l_m} \tag{18.24}$$

$$J_{ethanol} = -D_{em}\frac{\Delta Ce_m}{l_m} \tag{18.25}$$

Interfacial equilibrium at the membrane surface depends equity of the chemical potential. In general the chemical potential of the species in solution can be expressed as

$$\mu = \mu_o + RT\ln a, \tag{18.26}$$

where, μ_o is the chemical potential of pure permeant at $P = P_{ref}$

Thermodynamic equilibrium at both sides of the membrane for the species water and ethanol can be written as,

$$a_{wm1} = a_{w1}\exp[\frac{-v_w(P_{m1} - P_1)}{RT}] \tag{18.27}$$

$$a_{wm2} = a_{w2}\exp[\frac{-v_w(P_{m2} - P_2)}{RT}] \tag{18.28}$$

$$a_{em1} = a_{e1}\exp[\frac{-v_e(P_{m1} - P_1)}{RT}] \tag{18.29}$$

$$a_{em2} = a_{e2}\exp[\frac{-v_e(P_{m2} - P_2)}{RT}] \tag{18.30}$$

With the assumptions that pressure applied to the feed solution P_1 prevails across the membrane and the pressure decreases to that of permeate at the membrane permeate boundary. Hence we can write the flux terms as follows,

$$J_w = \frac{D_{wm}}{l_m}[C_{w1}K_{w1} - C_{w2}K_{w2}\exp[\frac{-v_w(P_1 - P_2)}{RT}]] \tag{18.31}$$

$$J_e = \frac{D_{em}}{l_m}[C_{e1}K_{e1} - C_{e2}K_{e2}\exp[\frac{-v_e(P_1 - P_2)}{RT}] \tag{18.32}$$

18.3   EXPERIMENTAL

Pervaporation was carried out for separation of ethanol and water separation in a laboratory scale Pervaporation setup supplied by Shivam Membrane, Pvt Ltd. A vacuum of less than 760 mmHg was generated with a vacuum pump. A condenser was used to condense the permeate. Recirculation pump was used to circulate the retentate. For Pervaporation experiments the flat sheet membrane module was used. The effective membrane area used for experimentation is 0.026 m². The schematic representation of Pervaporation process is shown in Figure 18.4.

The Pervaporation separations of ethanol–water mixtures were carried out over the different compositions at temperatures of 45, 55, 60, and 75°C, using PVA membrane. The analysis of permeate was done by Karl Fischer Titration. The PVA membrane is hydrophilic, thus increasing concentration of the ethanol in feed, it decreases the sorption of feed liquid by the membrane, resulting in a reduction of polymer swelling. Effect of temperature and concentration of the flux of membrane was studied and compared with model values.

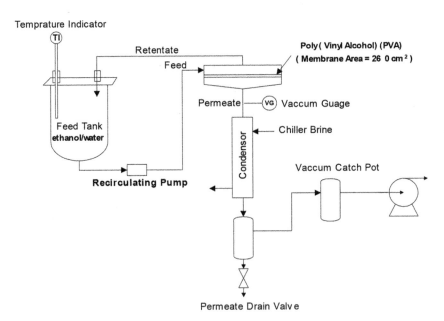

FIGURE 18.4   Experimental setup of pervaporation process.

18.4   RESULT AND DISCUSSION

The comparison of experimental and model values of total flux and selectivity are shown in Table 18.1. A sorption phenomenon at the liquid membrane interface is affected due to change in the feed composition. This can be proved by the solution-diffusion principle. The diffusion of the components in the membrane is dependent on the concentration of the components, hence the Pervaporation characteristic too is dependent on the same.

Figure 18.5a–c shows the effect of feed concentration on flux of membrane. It is observed that the flux of membrane decreases with decrease in temperature. The graphs were plotted for both experimental and model data. Good agreement can be seen between experimental flux and model flux data. Figure 18.6e–g shows the effect of feed concentration on selectivity; here selectivity decreases with increase in temperature.

TABLE 18.1   Comparison of Experimental and Model Values of Total Flux and Selectivity.

Temp. (°C)	Water content of feed (wt%)	Experimental data		Model data	
		Total flux (gm/ m² h)	Selectivity	Total flux (gm/m² h)	Selectivity
75	6.25	100.71	316.67	100.66	316.72
60		85.15	443.91	79.81	442.91
55		52.65	562.96	58.84	563.71
45		21.89	1143.39	22.69	1143.44
75	10.33	111.84	151.25	111.81	151.92
60		96.12	321.20	92.12	307.80
55		72.54	302.50	77.10	312.55
45		38.97	660.00	39.55	660.67
75	14.34	112.11	126.44	112.31	126.11
60		100.47	191.81	96.77	198.47
55		78.95	227.14	81.63	222.15
45		40.25	469.43	40.41	469.10

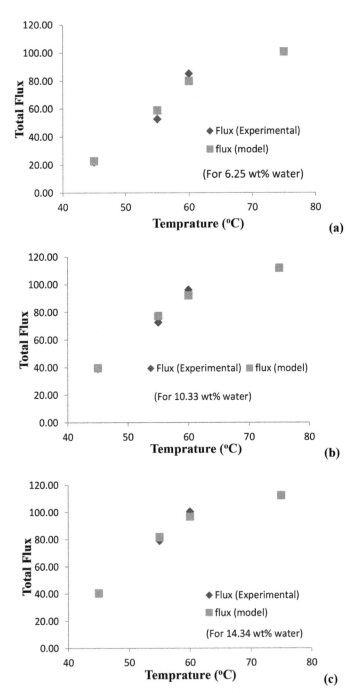

FIGURE 18.5a–c   Effect of feed composition on total flux.

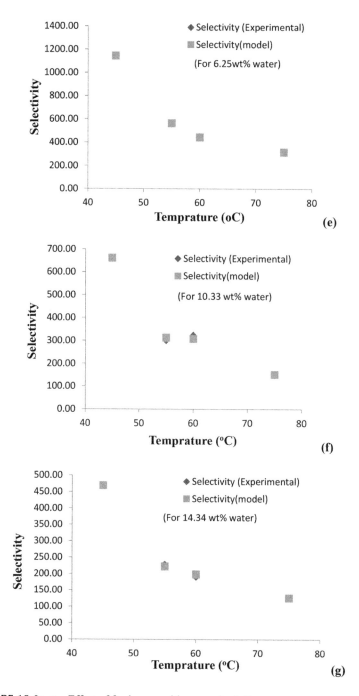

FIGURE 18.6e–g   Effect of feed composition on selectivity.

18.5   CONCLUSION

Pervaporation separation of an ethanol–water mixture through PVA membrane is a process which is described by several factors. In this study mathematical model was developed based on One Pass Pervaporation System to describe the behavior of Pervaporation process. The fluxes of the component ethanol and water were investigated over the membrane thickness. The interfacial concentrations at both the sides of membrane interface were found using thermodynamic model. The results estimated using the mathematical models are in good agreement with the experimental data.

KEYWORDS

- **pervaporation**
- **membrane**
- **flux**
- **selectivity**
- **modeling**

REFERENCES

1. Baker, R.W. *Pervaporation Membrane Technology and Applications;* 2nd ed.; Wiley: New York, 2004; pp 355–392.
2. Chen, J. H.; Liu, Q. L.; Zhang, X. H.; Zhang, Q. G. Pervaporation and Characterization of Chitosan Membranes Cross-Linked by 3-Aminopropyltriethoxysilane. *J. Membr. Sci.* **2007,** *292,* 125–132.
3. Huang, R. Y. M.; Shao, P.; Feng, X.; Anderson, W. A. Separation of Ethylene Glycol–Water Mixtures Using Sulfonated Poly(Ether Ether Ketone) Pervaporation Membranes: Membrane Relaxation and Separation Performance Analysis. *Ind. Eng. Chem. Res.* **2002,** *41,* 2957–2965.
4. Aptel, P.; Challard, N.; Cuny, J.; Neel, J. Application of the Pervaporation Process to Separate Azeotropic Mixtures. *J. Membr. Sci.* **1976,** *1,* 271–287.
5. Semenova, S. I.; Ohya, H.; Soontarapa, K. Hydrophilic Membranes for Pervaporation: An Analytical Review. *Desalination.* **1997,** *110,* 251–286.
6. Huang, R. Y. M.; Lm, V. J. C. Separation of Liquid Mixtures by Using Polymer Membranes I Permeation of Binary Organic Liquid Mixtures through Polyethylene. *J. Appl. Polym. Sci.* **1968,** *12,* 2615–2631.
7. Kim, S. N.; Kammermeyer, K. Actual Concentration Profiles in Membrane Permeation. *Sep. Sci.* **1970,** *5,* 679–697.

8. Lee, C. H. Theory of Reverse Osmosis and Some Other Membrane Permeation Operations. *J. Appl. Polym. Sci.* **1975,** *19,* 83–95.

9. Fels, M.; Huang, R. Y. M. Theoretical Interpretation of the Effect of Mixtures Composition on Separation Liquid M Polymers. *J. Macromol. Sci. Phys.* **1971,** *B5,* 89–109.

10. Fels, M.; Huang, R. Y. M. Diffusion Coefficients of Liquids in Polymer Membranes by a Adsorption Method. *J. Appl. Polym. Sci.* **1970,** *14,* 523–536.

11. Mulder, M. H. V.; Smolders, C. A. On the Mechanism of Separation of Ethanol/Water Mixtures by Pervaporation I Calculations of Concentration Profiles. *J. Membr. Sci.* **1984,** *17,* 289–307.

12. Mulder, M. H. V.; Franken, A. C. M.; Smolders, C. A. On the Mechanism of Separation of Ethanol Water Mixtures by Pervaporation II Experimental Concentration Profiles. *J. Membr. Sci.* **1985,** *23,* 41–58.

CHAPTER 19

MODELING AND SIMULATION OF CRYSTALLIZER USING COPPER OXIDE BASED NANOFLUIDS

G. P. LAKHAWAT[1] and R. P. UGWEKAR[2,*]

[1]Department of Chemical Engineering, Priyadarshini Institute of Engineering and Technology, Nagpur 440019, Maharashtra, India

[2]Department of Chemical Engineering, Laxminarayan Institute of Technology, Nagpur 440033, Maharashtra, India

*Corresponding author. E-mail: ugwekar@gmail.com

CONTENTS

ABSTRACT

New strategies for industrial world have to be developed to improve the thermal behavior of fluids used in crystallizer for energy conservation and thermal management. Generally cold water is used to cool the crystallizer so as to get product in crystal form. This investigation on crystallization operation by adding three different volume concentrations 1.0, 2, and 3% (V/V) of copper oxide nanoparticles in base fluid (distilled water) of size 45.46 nm synthesized by chemical precipitation method. The effect of nonionic surfactants Rokanol K7 (500 ppm) on crystallization operation was also investigated and it was observed that the heat transfer coefficient increases rapidly with surfactants. Maximum enhancements in the heat transfer were achieved using about 3 wt % copper oxide based nanofluids; however, a decreasing variation was observed at higher concentrations. The mathematical model has been developed by dimensional analysis followed by interpretation and the quantitative analysis of the model. For the objective of modeling, the experimental results obtained from crystallizer were validated with mathematical modeling for calculation of reliability of model. The reliability of mathematical model is obtained as 82.72, 83.75, and 86.51% for 1, 2, and 3% nanofluids, respectively, which shows its indicator of performance.

19.1 INTRODUCTION

Any physical phenomenon is a time being process and it has some causes and effects for its occurrence. If one forms the quantitative relationship between these causes and effects, then it can be considered as mathematical model. However, for deciding the causes and effects one has to understand the physics involved in the phenomenon.[1] The present phenomenon, that is, enhancement heat transfer coefficient of crystallizer, is a complex one and it cannot be reasoned out by any logic-based model. Further, it is also quite difficult to reason out by a field data based model, because certain variables cannot be changed as per our wish. The following sections throw some light on the process of model formulation, interpretation of model, and the quantitative analysis of mathematical model by calculating its reliability.

19.2 IDENTIFICATION OF VARIABLES

The genesis of model formulation takes place with the identification of variables for phenomenon. These variables fall under three categories: namely, (a) independent variables or causes, (b) dependent variables or effects, and (c) extraneous variable. Generally, the independent variables are those, which can vary according to the desire of experimenter, without giving any variation to other variables of the phenomenon. While dependent variable are those that only vary when the variation in the independent variables takes place. The extraneous variables are those that vary randomly and in an uncontrolled manner; indeed there is no control of experimenter on extraneous variables.[1] In crystallization operation, some prominent variables have been identified (Table 19.1).

TABLE 19.1 List of Independent/Dependent Variables of Crystallization Operation.

S. N.	Dependent/independent variable	Name of variable	MLT form	Representation
1	Dependent variable	Heat transfer coefficient	$H\,\theta^{-1}\,L^{-2}\,T^{-1}$	H
2	Independent variable	Thermal conductivity of nanofluids	$H\,\theta^{-1}\,L^{-1}\,T^{-1}$	K
3	Independent variable	Viscosity of nanofluids	$M\,L^{-1}\,\theta^{-1}$	μ
4	Independent variable	Density of nanofluids	ML^{-3}	P
5	Independent variable	Specific heat of nanofluids	$HM^{-1}T^{-1}$	Cp
6	Independent variable	Velocity	$L\theta^{-1}$	U
7	Independent variable	Temp diff of cooling	T	ΔT
8	Independent variable	Speed (rpm)	θ^{-1}	N
9	Independent variable	Particle size	L	dp
10	Independent variable	Conc. of solute	ML^{-3}	Cs

Reduction of variables through dimensional analysis:

$$h = f[K^a\,\mu^b\,P^c\,Cp^d\,u^e\,\Delta T^f\,n^g\,d_p^{\,i}\,Cs^j]$$

The resultant equation is

$$h = f[(K)^a(\mu)^b(P)^c(Cp)^{1-a}(u)^e(\Delta T)^0(n)^{1-a-b-e}(d_p)^{1-2a-2b-e}(Cs)^{1-a-b-c}]$$

Now by simplifying one will get

$$\left[\left(\frac{h}{Cp.n.d_p.Cs}\right)\right] = \varphi \times \left[\left(\frac{k}{Cp.n.d_p^2.Cs}\right)\right]^a \left[\left(\frac{\mu}{n.d_p^2.Cs}\right)\right]^b \left[\left(\frac{\rho}{Cs}\right)\right]^c \left[\left(\frac{u}{n.d_p}\right)\right]^e \quad (19.1)$$

19.3 EXPERIMENTATION PROCEDURE AND DATA GENERATION

As the comparison is between the performances of crystallizer with nano-fluid (CuO/water) and the base fluid (water),[2] first the performance of crys-tallizer with base fluid (water) and then with nanofluids were checked. The initial temperatures of inlet of hot solution and cold fluid are constant, but later inlet and outlet temperatures of hot solution and cold fluid vary throughout the experiment and the flow rate of hot fluid is also constant. The study of variation in performance of crystallizer with different flow rate of cold fluid (nanofluid and base fluid) and volumetric concentration of nano-fluid has been done. While doing this the inlet temperature of hot solution in the heating tank is maintained at 60°C and the inlet temperature of cold fluid is maintained at 25°C. As the cold fluid is recycled through same tank, its inlet and outlet temperatures vary gradually. The inlet temperature of cold fluid goes on increasing while outlet temperature goes on deceasing. The hot solution inside the tank is agitated slowly at 1.3 rps. The flow rate of cold fluid is set.[3] First of all, a hot solution of magnesium sulfate is introduced in the inner tank and heater is started to attain the temperature of 60°C with constant agitation for a few minutes, so that the wall of inner tank attains the constant temperature of hot solution. Then the pump is started to flow the cold fluid through the jacket. As soon as the cold fluid starts flowing the hot fluid starts cooling with high rate and temperature of all the three, that is, inlet temperature of cold fluid, outlet temperature of cold fluid, and tempera-ture of solution, are measured by temperature indicator of control system. Note down the temperatures of both the hot and cold fluids. After efficient cooling of hot solution of magnesium sulfate for several hours, the crystals of magnesium sulfate start forming. When the temperature of hot solution of magnesium sulfate is reached its minimum value, the crystals are formed. The mother liquor is discharged from the outlet provided at the bottom of tank. The supernatant solution of mother liquor is discarded carefully. The crystals of magnesium sulfate are filtered and then heated in oven at 50°C, the anhydrous powder of magnesium sulfate is obtained. Repeat the proce-dure for different flow rates and compare the heat transfer coefficients of base fluids used.[4] For studying the performance of crystallizer using nano-fluid as a coolant, the same above procedure in the same inlet temperature of hot and cold fluids is repeated just by replacing the base fluid (water) by adding three different volume concentrations 1, 2, and 3% (v/v) of copper oxide nanoparticles in base fluid (Tables 19.2 and 19.3), and the readings of temperature of hot solution and inlet and outlet temperatures of nanofluid are noted down for seven different flow rates. Further the effect of surfactant loading was also studied on crystallization process which is reported in Tables 19.4 and 19.5.

TABLE 19.2 Effect of Base Fluids and 1% Nanofluid Flow Rate on Crystallization ($T_{sol\ in}$ = 60°C and $T_{cold\ in}$ = 25°C).

Sr. No.	Cold fluid flow rate (l/h)	$T_{sol\ out}$	$T_{cold\ out}$	h_o	U_o	$T_{sol\ out}$	$T_{cold\ out}$	h_o	U_o
		Cold water				1% copper oxide			
1	20	42.25	44.25	301	165	36.25	49.25	316	178
2	30	41.25	45.27	333	187	35.41	50.42	348	199
3	40	40.23	46.21	356	201	34.27	51.47	371	212
4	50	39.42	47.52	386	223	33.91	52.28	399	235
5	60	37.85	49.78	418	243	32.76	53.16	439	257
6	70	36.45	50.14	448	255	31.14	54.37	465	267
7	80	35.25	51.23	470	273	30.13	55.44	485	285

TABLE 19.3 Effect of 2 and 3% Nanofluid Flow Rate on Crystallization ($T_{sol\ in}$ = 60°C and $T_{cold\ in}$ = 25°C).

Sr. No.	Cold fluid flow rate (l/h)	$T_{sol\ out}$	$T_{cold\ out}$	h_o	U_o	$T_{sol\ out}$	$T_{cold\ out}$	h_o	U_o
		2% copper oxide				3% copper oxide			
1	20	34.25	51.31	328	185	33.23	53.25	336	192
2	30	33.14	52.42	359	210	32.52	54.14	367	218
3	40	32.26	53.28	383	223	31.21	53.26	387	231
4	50	31.45	54.19	409	246	30.49	54.18	417	252
5	60	30.42	55.27	452	269	29.27	55.16	461	278
6	70	29.74	56.52	477	279	28.29	56.27	485	286
7	80	28.34	56.88	497	296	27.43	57.12	504	304

TABLE 19.4   Effect of Surfactant (1% Copper Oxide + RK7) on Crystallization ($T_{sol\ in}$ = 60°C and $T_{cold\ in}$ = 25°C).

Sr. No.	Cold fluid flow rate (l/h)	$T_{sol\ out}$	$T_{cold\ out}$	h_o	U_o	$T_{sol\ out}$	$T_{cold\ out}$	h_o	U_o
		Cold water				(1% copper oxide + RK7)			
1	20	42.25	44.25	301	165	34.23	50.12	322	184
2	30	41.25	45.27	333	187	33.24	51.24	354	205
3	40	40.23	46.21	356	201	32.28	52.28	377	218
4	50	39.42	47.52	386	223	31.43	53.26	405	242
5	60	37.85	49.78	418	243	30.62	54.18	445	263
6	70	36.45	50.14	448	255	29.37	55.16	470	272
7	80	35.25	51.23	470	273	28.14	56.45	490	291

TABLE 19.5   Effect of Surfactant (2 and 3% Copper Oxide + RK7) on Crystallization ($T_{sol\ in}$ = 60°C and $T_{cold\ in}$ = 25°C).

Sr. No.	Cold fluid flow rate (l/h)	$T_{sol\ out}$	$T_{cold\ out}$	h_o	U_o	$T_{sol\ out}$	$T_{cold\ out}$	h_o	U_o
		(2% copper oxide + RK7)				(3% copper oxide + RK7)			
1	20	33.41	52.24	334	190	32.23	53.14	342	197
2	30	32.19	53.52	364	216	31.63	54.56	373	224
3	40	31.31	54.19	387	229	30.21	55.29	392	237
4	50	30.28	55.75	416	251	29.52	56.13	423	257
5	60	29.42	56.38	458	275	28.37	57.66	465	284
6	70	28.63	57.18	483	285	27.42	58.43	492	292
7	80	27.11	58.28	503	302	26.46	59.22	510	310

The indices of different π terms aimed at model can be identified by using multiple regression analysis[5,6]. By considering three independent π terms and one dependent π term, let the model aimed at be of the form,

$$[\pi] = \phi \ [\pi_1]^a \ [\pi_2]^b \ [\pi_3]^c \ [\pi_4]^e,$$

Solving the constant values a, b, c, and e by regression methods, we get

$\phi = 0.9645$, $a = 0.6517$, $b = -0.1247$, $c = -0.2039$, $e = 0.7815$.

By substituting above values one can get following model,

$$[\pi] = 0.9645 \, [\pi_1]^{0.6517} \, [\pi_2]^{-0.1247} [\pi_3]^{-0.2039} [\pi_4]^{0.7815}.$$

For 1, 2, and 3% copper oxide based nanofluids, the model becomes

$$\left[\left(\frac{h}{Cp.n.d_p.Cs}\right)\right] = 0.9645 \times \left[\left(\frac{k}{Cp.n.d_p{}^2.Cs}\right)\right]^{0.6517} \left[\left(\frac{\mu}{n.d_p{}^2.Cs}\right)\right]^{-0.1247} \left[\left(\frac{\rho}{Cs}\right)\right]^{-0.2039} \left[\left(\frac{u}{n.d_p}\right)\right]^{0.7815} \quad (19.2)$$

$$\left[\left(\frac{h}{Cp.n.d_p.Cs}\right)\right] = 0.9645 \times \left[\left(\frac{k}{Cp.n.d_p{}^2.Cs}\right)\right]^{0.69552} \left[\left(\frac{\mu}{n.d_p{}^2.Cs}\right)\right]^{-0.1438} \left[\left(\frac{\rho}{Cs}\right)\right]^{-0.2562} \left[\left(\frac{u}{n.d_p}\right)\right]^{0.8229} \quad (19.3)$$

$$\left[\left(\frac{h}{Cp.n.d_p.Cs}\right)\right] = 0.9645 \times \left[\left(\frac{k}{Cp.n.d_p{}^2.Cs}\right)\right]^{0.7322} \left[\left(\frac{\mu}{n.d_p{}^2.Cs}\right)\right]^{-0.1978} \left[\left(\frac{\rho}{Cs}\right)\right]^{-0.3166} \left[\left(\frac{u}{n.d_p}\right)\right]^{0.8426} \quad (19.4)$$

19.4 INTERPRETATION OF MODEL

Equation 19.2 is established based on experimental data carried out during experimentation. It indicates that π_4 term, which relates to flow properties of experimental set-up, has the highest influence as 0.7815 on effect, that is, heat transfer coefficient, while the least influence is seen for π_2 as 0.1247, which relates to viscous property of experimental set-up. The π_1 and π_3 related thermal property and density properties have moderate influence as 0.6517 and −0.2039, respectively, on the effect.

19.5 RELIABILITY OF MODEL

The reliability is a term which indicates the performance of a mathematical model. In this study, reliability of model is estimated as follows. With reference to the model, the known values of independent π terms have been submitted in the model and thus obtained the required dependent variable; generally it is called calculated values of dependent variables. Now, one could find the error by subtracting the calculated values from observed values of dependent variable. Once the error is estimated, then reliability can be estimated by calculating the mean square error.

This can be done by using following formula,

Reliability = 1 − mean square error

$$\text{mean square error} = \sum_{i=1}^{n} \left(\text{HTC, cal} - \text{HTC, obs}\right)^2,$$

where HTC, cal is the heat transfer coefficient based on calculated value (model value) and HTC, obs is the heat transfer coefficient based on observed value (experimental value).Hence for the present model reliability for 1, 2, and 3% copper oxide based nanofluids are obtained as 82.72, 83.75, and 86.51%, respectively. As shown in Figure 19.1, the reliability of the model seems to increase in percent, so it indicates that experimental results were satisfactorily validated with modeling result.

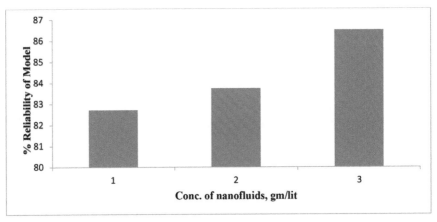

FIGURE 19.1   Reliability of model at different concentration of nanofluids.

19.6   CONCLUSION

A batch crystallization of magnesium sulfate was done in this experimentation. Heat transfer coefficient is higher with nanofluids than without nanofluids. In this mathematical model, it is seen that the variables related to flow properties have the highest influence and the viscous properties have less influence on heat transfer coefficient. The quantitative analysis of the models gives some important conclusions that the reliability values of mathematical model for 1, 2, and 3% copper oxide based nanofluids are obtained as 82.72, 83.75, and 86.51%, respectively. Maximum enhancements in the reliability of model were achieved using about 3 wt% copper oxide nanoparticles; however, a decreasing variation was observed at higher concentrations. The microconvection and particle aggregation due to the interpenetration layers can provide this kind of variation.

KEYWORDS

- crystallizer
- nanofluid
- surfactant
- thermal management
- reliability

REFERENCES

1. Schenck, H. Jr. *Theories of Engineering Experimentation;* 1st ed.; McGraw Hill Inc.: New York, 1967.
2. Abbasia, M.; Kermanpura, A.; Emadia, R. Effects of Thermo-Mechanical Processing on the Mechanical Properties and Shape Recovery of the Nanostructured Ti50Ni45Cu5 Shape Memory Alloy. *Procedia Mater. Sci.* **2015**, *11,* 61–66.
3. Ugwekar, R.; Lakhawat, G. Influence of Nanofluid on the Performance of Crystallization. *Chem. Sci. Trans.* **2014**, *3,* 614–623.
4. Paul, E.; Tung, H. Organic Crystallization Processes. *Powder Technol.* **2005**, *150,* 133–143.
5. Ugwekar, R ; Lakhawat, G. Formulation of approximate generalized experimental data based model in a liquid—liquid extraction using zinc oxide nanomaterial. *Perspectives in Science* **2016**, *8,* 664—666.
6. Lakhawat, G; Ugwekar, R. Effect of nonionic surfactant additives on the performance of nanofluids in heat exchanger. *Int. J Nano Dimension* **2017**, *8,* 18-30.

CHAPTER 20

SENSITIVITY ANALYSIS OF SHELL AND TUBE HEAT EXCHANGER USING CHEMCAD

S. M. WAGH, D. P. BARAI*, and M. H. TALWEKAR

Laxminarayan Institute of Technology, Rashtrasant Tukadoji Maharaj Nagpur University, Nagpur, Maharashtra, India

*Corresponding author. E-mail: divyapbarai@yahoo.co.in

CONTENTS

ABSTRACT

Shell and tube heat exchangers are used extensively throughout the process industry. The inlet conditions of the fluids on the shell side or on the tube side have an impact on the working of the heat exchanger. Designers and engineers study the sensitivity analysis to understand and control the system for the various effects of operating and geometric parameters. In this chapter, the sensitivity of outlet parameters like the hot water outlet temperature and enthalpy, overall heat transfer coefficient, and heat duty to the various inlet parameters are studied using CHEMCAD. Thus, the values of different outlet operating parameters can be known, as the inlet operating parameters are changed.

20.1   INTRODUCTION

Sensitivity is defined as a technique for understanding the impact on system performance or by changing various input values about which there is uncertainty.[1] The different simulation outputs in thermal design are studied for the impact of various operating and geometric parameters. However, the analysis of thermal models is more complex because of the high number of input parameters. The goal of sensitivity analysis is to understand the changes in magnitude of different output variables due to impact of minor changes in operating parameters. Modern engineering designs have to make extensive use of computer models for testing before they are manufactured. Systems are to be analyzed for the sources of different uncertainties to design and engineer robust models. Such models are to be tested and verified for the robustness of various systems and for the presence of different uncertainties.[2]

In a shell and tube heat exchanger, two same or different fluids at different temperatures for heat transfer are passed through the shell and the tubes for heat transfer between them. Thus, their flow rates and inlet temperatures seem to be affecting the outlet parameters of the exchanger. Parallel flow heat exchanger has been studied by Osueke et al.[3] for the effects of fluid flow rate. Similar studies for a corrugated plate type heat exchanger have also been done by Murugesan and Balasubramanian.[4] The variations of log mean temperature difference, Nusselt number, Prandel number, heat transfer coefficient, and pressure drop with change in values of Reynolds number for shell and tube heat exchanger at different orientations are studied by Singh and Sehgal.[5] A study on pressure drop based on changing mass flow rate in a plate-fin heat exchanger is conducted by Asadi and Khoshkhoo.[6] Analysis of a parallel flow double pipe heat exchanger for varied mass flow conditions both in inner and outer tubes has been carried out by Kumar et al.[7]

The cold water flow rate was varied in a range from 1000 to 30,000 kg/h at different cold water inlet temperatures as well as hot water flow rate in the range from 1000 to 30,000 kg/h at various hot water inlet temperatures. Temperature range considered for cold water inlet is 273–323 K and that considered for hot water is 323–373 K. By varying the cold water flow rate, changes in the values of hot water outlet temperature and enthalpy, overall heat transfer coefficient, and heat duty were noted. Curves are drawn each for different cold water inlet temperatures. By changing hot water flow rate, the changes in the values of overall heat transfer coefficient, and heat duty of the exchanger were noted. Curves are drawn each for various hot water inlet temperatures.

The independent variables are needed to be specified. These are the parameters which will be varied over a user-specified range in a specified number of steps. This variable drives the sensitivity analysis. Any two independent variables are permitted in any given sensitivity analysis. When these independent variables are varied, variables whose values are calculated and recorded are called the dependent variables. After running, the data can be noted and plotted.

20.2 PROCEDURE

In CHEMCAD, the sensitivity analysis for an equipment can be carried out by clicking Run>Sensitivity Study>New Analysis on the toolbar on the CHEMCAD interface. First, a name for the analysis has to be specified. In the Edit Sensitivity Menu, there is one tab named "Adjusting" and several other tabs named "Recording." The "Adjusting" tab enables you to specify the independent variables. The independent variables considered for the study are cold water flow rate, cold water inlet temperature, hot water flow rate, and hot water inlet temperature. Cold water flow rate and hot water flow rate are changed from 1000 to 30,000 kg/h. The range for cold water inlet temperature considered is from 273 to 323 K. And the range for hot water inlet temperature considered is from 323 to 373 K. The "Recording" tab enables you to specify the dependent variables. Dependent variables recorded in this study are hot water outlet temperature, hot water outlet enthalpy, overall heat transfer coefficient, and heat duty. After filling all the necessary data in both the tabs, this needs to be run. This is done by clicking Run>Sensitivity Study, and clicking on the name given to the analysis and then clicking on "Run All." After running, the data can be noted and plotted by clicking on "Plot Results" in the same menu.

20.3   RESULTS AND DISCUSSION

Figure 20.1 depicts the hot water outlet temperature against the cold water flow rate at constant hot water inlet flow rate of 12,000 kg/h at inlet temperature of 366 K. Various curves for different cold water inlet temperatures of 273, 283, 293, 303, 313, and 323 K are plotted. It is observed that as the cold water flow rate increases, the hot water outlet temperature decreases. Also, that the value of hot water outlet temperature attained at a particular flow rate is different for various cold water inlet temperatures. The flow rate of hot water being constant, the specific heat rate of the hot water remains almost constant. Thus, the amount of heat dissipated from the hot water is also constant. The heat lost by the hot fluid should be equal to the heat gained by the cold fluid according to the law of conservation of energy. Increase in cold water mass flow rate will lead to an increase in the specific heat rate of cold water, increasing the heat transfer. Thus, there is a decrease in the hot water outlet temperature. The curve for the temperature of 273 K of the cold water inlet is the lowermost one. It shows that the value of hot water outlet temperature that can be achieved at a particular flow rate of cold water at

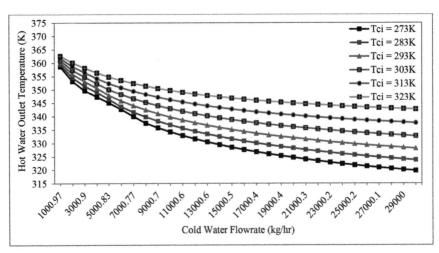

FIGURE 20.1   Effect of change in cold water flow rate on hot water outlet temperature for various cold water inlet temperatures.

273 K is less than that achieved at that particular flow rate of cold water at a temperature above 273 K. This is because the temperature difference at inlet between the cold water and hot water is maximum when cold water is at 273 K. Temperature difference, being the driving force for heat transfer, will cause the resultant outlet temperature of hot water to be lesser. Thus, it is seen that the hot water outlet temperature increases with increase in the cold water inlet temperature.

Figure 20.2 depicts the hot water outlet enthalpy against the cold water flow rate at constant hot water inlet flow rate of 12,000 kg/h at inlet temperature of 366 K. Various curves for different cold water inlet temperatures of 273, 283, 293, 303, 313, and 323 K are plotted for both cases. It is observed that as the cold water flow rate increases, the hot water outlet enthalpy decreases. Also, that the value of hot water outlet enthalpy attained at a particular flow rate is different for various cold water inlet temperatures. According to the laws in thermodynamics, the enthalpy is dependent on the temperature of a substance. Enthalpy of a substance is a function of its heat capacity and temperature. Heat capacity here in this case being constant, the enthalpy is directly proportional to the temperature. So, the enthalpy of hot water also decreases as the temperature decreases. Similar to the curves drawn for hot water outlet temperature, the curves for hot water outlet enthalpy are drawn each for different cold water inlet temperatures. Thus, it is seen that the hot water outlet enthalpy increases with increase in the cold water inlet temperature.

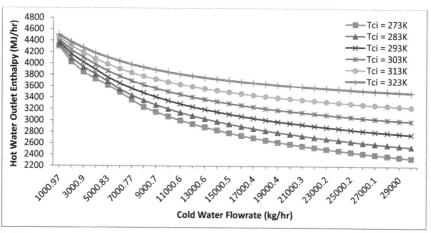

FIGURE 20.2 Effect of change in cold water flow rate on hot water outlet enthalpy at different cold water inlet temperatures.

Figure 20.3 depicts the overall heat transfer coefficient against the cold water inlet flow rate at constant hot water flow rate of 12,000 kg/h at inlet temperature of 366 K. Here also, the curves for various cold water inlet temperatures of 273, 283, 293, 303, 313, and 323 K are plotted. It is observed that the overall heat transfer coefficient increases with increase in the cold water inlet flow rate. Also, the value of overall heat transfer coefficient attained at a particular flow rate of cold water is different for various cold water inlet temperatures. Increase in cold water inlet flow rate on the tube side increases the velocity on the tube side. This results in an increase in the turbulence of the fluid which gives rise to higher film coefficient on the tube side and thus a higher overall heat transfer coefficient. The curve for the temperature of 323 K of the cold water inlet is the uppermost one. It shows that the value of overall heat transfer coefficient that can be achieved at a particular flow rate of cold water at 323 K is more than that achieved at that particular flow rate of cold water at a temperature below 323 K. This is because the logarithmic mean temperature difference is less when the temperature of cold water is more. This makes the overall heat transfer coefficient to be more. Thus, the overall heat transfer coefficient increases with increase in cold water temperature.

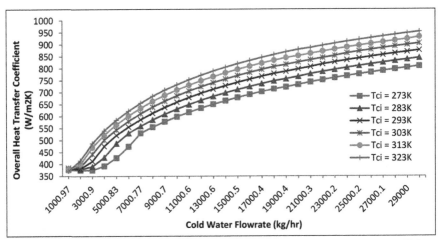

FIGURE 20.3   Effect of change in cold water flow rate on overall heat transfer coefficient at different cold water inlet temperatures.

Figure 20.4 depicts the overall heat transfer coefficient against the hot water inlet flow rate at constant cold water flow rate of 17,197 kg/h at inlet temperature of 288 K. Here also, the curves for different hot water inlet temperatures of 323, 333, 343, 353, 363, and 373 K are plotted. An increase in the overall heat transfer coefficient is observed as the hot water inlet flow rate increases. Also, the value of overall heat transfer coefficient attained at a particular flow rate of hot water is different for various hot water inlet temperatures. The increase in hot water inlet flow rate on the shell side increases the velocity on the shell side. This results in an increase in the turbulence of the fluid which gives rise to higher film coefficient on the shell side and thus a higher overall heat transfer coefficient. The curve for the temperature of 373 K of the hot water inlet is the uppermost one. It shows that the value of overall heat transfer coefficient that can be achieved at a particular flow rate of hot water at 373 K is more than that achieved at that particular flow rate of hot water at a temperature below 373 K. This is because the logarithmic mean temperature difference is less when the temperature of hot water is more. This raises the value of overall heat transfer coefficient. Thus, with the increase in hot water inlet temperature, overall heat transfer coefficient increases.

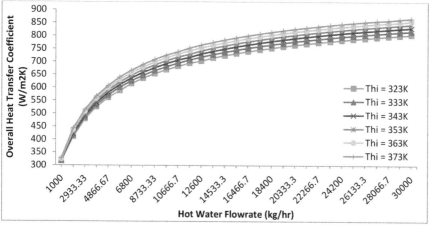

FIGURE 20.4   Effect of change in hot water flow rate on overall heat transfer coefficient at different hot water inlet temperatures.

Figure 20.5 depicts the heat duty of the exchanger against the cold water inlet flow rate at constant hot water flow rate of 12,000 kg/h at inlet temperature of 366 K. Here also, the curves for different cold water inlet temperatures of 273, 283, 293, 303, 313, and 323 K are plotted. With increase in the cold water inlet flow rate, an increase in the exchanger heat duty is observed. Also, the value of heat duty at a particular flow rate of cold water is different for different cold water inlet temperatures. The amount of heat exchanged between hot and cold fluids per unit time is the heat duty. It is a product of mass flow rate, temperature difference between the outlet and inlet of any of a fluid in the heat exchanger, and specific heat. The heat duty of the exchanger increases with increase in cold water flow rate. The curve for the temperature of 273 K of the cold water inlet is the uppermost one. It shows that the value of heat duty of the heat exchanger at a particular flow rate of cold water at 273 K is more than that achieved at that particular flow rate of cold water at a temperature above 273 K. The temperature difference between the outlet and inlet of cold water is maximum when the cold water is fed at 273 K. Heat duty increases with the increase in outlet and inlet temperature difference of cold fluid. So, the heat duty increases with decrease in cold water inlet temperature.

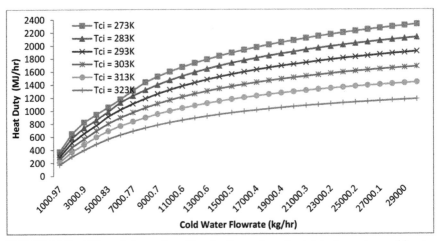

FIGURE 20.5   Effect of change in cold water flow rate on heat duty for various cold water inlet temperatures.

Figure 20.6 depicts the heat duty of the exchanger against the hot water inlet flow rate at constant cold water flow rate of 17,197 kg/h at inlet temperature of 288 K. Here also, the curves for various hot water inlet temperatures of 323, 333, 343, 353, 363, and 373 K are plotted. It is observed that the heat duty increases with increase in the hot water inlet flow rate. Also, the value of heat duty at a particular flow rate of hot water is different for various hot water inlet temperatures. The amount of heat exchanged between hot and cold fluids per unit time is the heat duty. It is a product of mass flow rate, temperature difference between the outlet and inlet of any of a fluid in the heat exchanger, and specific heat. So, increase in hot fluid flow rate increases the heat duty of the exchanger. The curve for the temperature of 373 K of the hot fluid inlet is the uppermost one. It shows that the value of heat duty of the heat exchanger at a particular flow rate of hot water at 373 K is more than that achieved at that particular flow rate of hot water at a temperature below 373 K. The temperature difference between the hot water inlet and outlet is maximum when the hot water is fed at 373 K. Heat duty increases with the increases in temperature difference between the outlet and inlet of hot fluid. So, increase in inlet temperature of hot water increases the heat duty.

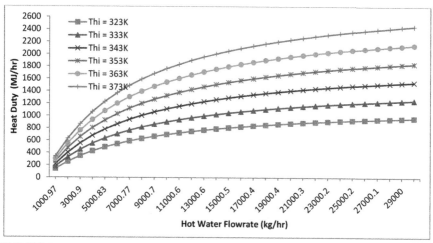

FIGURE 20.6   Effect of change in hot water flow rate on heat duty at different hot water inlet temperatures.

20.4   CONCLUSIONS

CHEMCAD software is used to simulate shell and tube heat exchanger to evaluate the performance. Sensitivity analysis for shell and tube heat exchanger was carried out for changes in few operating parameters and changes in other parameters were recorded. It is observed that the hot water outlet temperature and enthalpy decrease with increase in the cold water flow rate and inlet temperature. The overall heat transfer coefficient increases with increase in the cold water flow rate and inlet temperature and the hot water flow rate and inlet temperature. It is also observed that the exchanger heat duty increases with decrease in cold water inlet temperature and that it increases with increase in hot water inlet temperature.

KEYWORDS

- **sensitivity analysis**
- **shell and tube heat exchanger**
- **CHEMCAD**

REFERENCES

1. Marshall, H. E. *Sensitivity Analysis, Technology Management Handbook;* CRC Press LLC: Boca Raton, Florida, 1999.
2. Saltelli, A.; Tarantola, S.; Campolongo, F.; Ratto, M. *Sensitivity Analysis in Practice: A Guide to Assessing Scientific Models;* Wiley: New York, 2004.
3. Osueke, C. O.; Onokwai, A. O.; Adeoye, A. O. Experimental Investigation on the Effect of Fluid Flow Rate on the Performance of a Parallel Flow Heat Exchanger. *Int. J. Innov. Res. Adv. Eng.* **2015,** *2,* 10–23.
4. Murugesan, M. P.; Balasubramanian, R. The Effect of Mass Flow Rate on the Enhanced Heat Transfer Characteristics in a Corrugated Plate Type Heat Exchanger. *Res. J. Eng. Sci.* **2012,** *1,* 22–26.
5. Singh, A.; Sehgal, S. S. Thermohydraulic Analysis of Shell-and-Tube Heat Exchanger with Segmental Baffles. *ISRN Chem. Eng.* **2013,** *2013,* 1–5.
6. Asadi, M.; Khoshkhoo, R. H. Effects of Mass Flow Rate in Terms of Pressure Drop and Heat Transfer Characteristics. *Merit Res. J. Environ. Sci. Toxicol.* **2013,** *1,* 5–11.
7. Kumar, S.; Karanth, K. V.; Murthy, K. Numerical Study of Heat Transfer in a Finned Double Pipe Heat Exchanger. *WJMS.* **2015,** *11,* 43–54.

CHAPTER 21

EXTRACTIVE DISTILLATION SIMULATION FOR THE SEPARATION OF METHYLCYCLOHEXANE AND TOLUENE MIXTURE WITH PHENOL AS AN EXTRACTOR USING CHEMCAD

S. S. ANWANI and S. P. SHIRSAT*

Laxminarayan Institute of Technology, Rashtrasant Tukadoji Maharaj Nagpur University, Nagpur 440033, Maharashtra, India

Corresponding author. E-mail: spslit@rediffmail.com

CONTENTS

ABSTRACT

Methylcyclohexane (MCH) and toluene are close boiling point components and pose a huge bottleneck in the distillation process. The better practice is to use the high boiling component phenol as an extractive agent which increases the volatility of MCH with respect to toluene and thereby reduces the number of stages required for desired degree of purity than if phenol would not have been used. In this steady-state simulation we have simulated the process in CHEMCAD software (6.3.1) and tried to see the effect of changing the flow rate of the extractive solvent which gives criterion about the optimum flow rate in the distillation column. In this chapter we have simulated the effect of reflux ratio and the number of stages on the purity of MCH and studied the variation in reboiler heat duty.

21.1   INTRODUCTION

Distillation is a method of separating mixtures based on differences in their volatilities in a boiling liquid mixture.[1] But in the case of the extractive distillation, the process of simple distillation is not useful. To separate the close boiling components, different distillation methods can be used such as extractive distillation, membrane distillation, or pressure swing distillation. For the separation of methylcyclohexane (MCH) and toluene mixture, simulation was carried out by using CHEMCAD software (6.3.1).

Extractive distillation is the method of separating close boiling components in the presence of some suitable solvents or a mixture of solvents. In the case of separation of MCH and toluene, the solvent used is phenol.[2-4] The boiling point of MCH is 101.8°C and that of toluene is 110°C. As a result of the low boiling point difference they are very hard to separate, thereby it has become necessary to use the third component which is phenol in this case. Phenol is a fairly non-volatile component that has a boiling point of 180°C, which is very high with respect to the above system. Phenol improves the volatility of MCH and thus helps to recover almost pure MCH in the overhead distillate and phenol and toluene mixture in the bottom of distillation column.

21.2   VAPOR LIQUID EQUILIBRIUM

Vapor liquid curve of MCH and toluene can be directly obtained by plot function in CHEMCAD (6.3.1). As we can see from the T–x–y plot (Fig. 21.1), upper curve and lower curves are very close, which is not desirable for the feasibility of the column, leads to increase in number of plates, and directly increases the cost of column. This indicates to use other techniques to be implemented such as extractive distillation using a third component for making this separation cost effective.

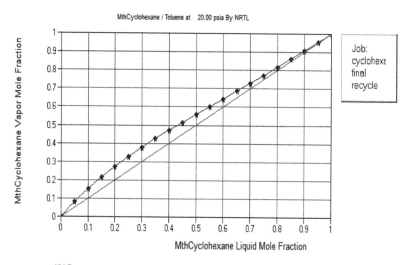

FIGURE 21.1   *T–x–y* curve at 20 psia for MCH and toluene.

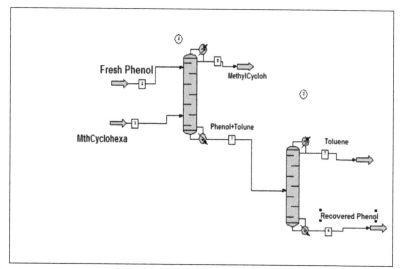

FIGURE 21.2   CHEMCAD flow diagram of extractive distillation.

21.3 PROCEDURE

MCH and toluene both at the rate of 200 lbmol/h each at 220°F and 20 psia pressure are fed to the 14th stage of the first distillation column. Fresh phenol having a molar flow rate of 1200 lbmol/h, along with 100 lbmol/h of recycled phenol from the solvent recovery column, is also charged into the seventh tray of the extractive distillation column. Extractive distillation column has 33 trays. To get the desired 99.5% purity of MCH, reboiler duty of 23,000 MJ/h, reflux ratio of 3.94, and 33 trays are specified.

In the solvent recovery column, a mixture of phenol and toluene (total flow rate = 1399 lbmol/h, 85.7 mol% phenol, and 14.1 mol% toluene) are fed to the middle tray of column having 30 trays. To achieve the maximum recovery of toluene (199.7 lbmol/h) from the top of the column, reflux ratio of 1.44, a reboiler duty of 10,000 MJ/h, and 30 trays are set. Modified UNIversal QUAsiChemical (UNIQUAC) models are used to run the simulation.

Phenol recovered from the bottom of the solvent recovery column may be sent to splitter so that majority of phenol can be purged for treatment to remove any impurities present. Only the 100 lbmol/h from the splitter is being sent to the solvent mixer from where it combines with fresh solvent, and mixed stream enters the column. The process simulation of the above system is shown in Figure 21.2. The temperature and concentration profiles have directly drawn using plot function in CHEMCAD (6.3.1), which is illustrated in Figures 21.3–21.7.

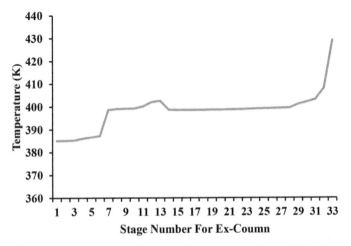

FIGURE 21.3   Tray temperature profile curve for extractive column (EX-column).

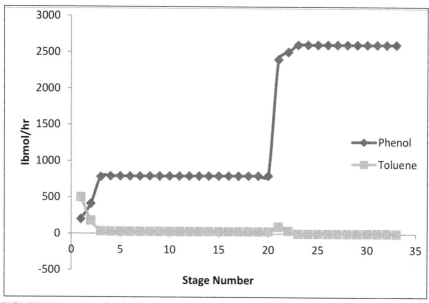

FIGURE 21.4   Tray liquid flow profile for solvent recovery column.

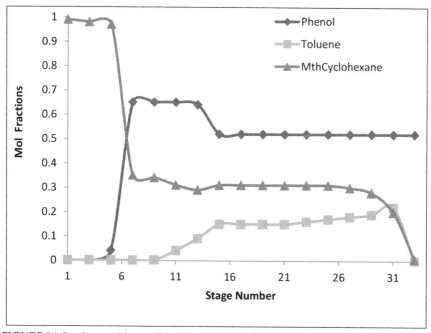

FIGURE 21.5   Composition profile for extractive column.

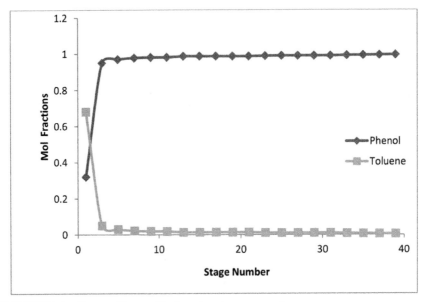

FIGURE 21.6   Composition profile for solvent recovery column.

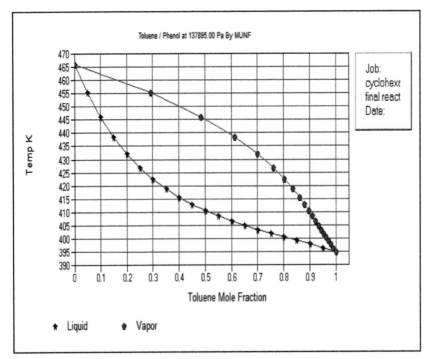

FIGURE 21.7   *T–x–y* plot for toluene-phenol system at 20 psia.

21.4 RESULTS AND DISCUSSION

The complete simulation was run using modified UNIQUAC activity coefficients and non-ideality of liquid phase was corrected by replacing component concentration with activities. Figure 21.8 shows the graph plotted as reflux ratio on x-axis and number of stages and reboiler duty of each column on y-axis. The condition is found out from the intersection of reboiler duty and the number of stages curves plotted with respect to reflux ratio. The effect of phenol flow rate on the overhead purity of MCH has also studied to ensure the optimal flow condition of solvent in the column. For the solvent recovery column the reflux ratio selected is 1.44 with the reboiler duty of 10,000 MJ/h and the number of trays set is 30, with feed tray location being 15. The MCH percent in the top of extractive distillation column increases as the flow rate of phenol rises; so 1200 lbmol/h had been selected because desired purity has already been achieved at this flow rate and no further change has seen beyond this flow rate.

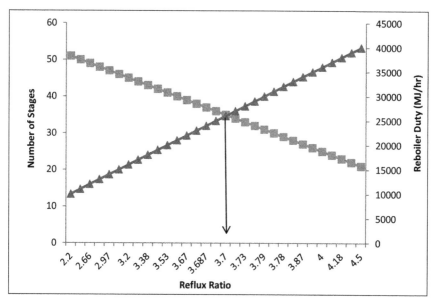

FIGURE 21.8 Optimum reflux ratio for ex-distillation column.

As shown in Figure 21.9, for solvent recovery column, reflux ratio is plotted on x-axis, while y-axis shows reboiler duty in MJ/h, by keeping the purity of toluene constant equal to 0.99% for a given number of plates, effect of reflux ratio on reboiler duty has been studied and found that as the reflux ratio increases the heat load on the reboiler increases. Minimum heat duty selected was 10,000 MJ/h, which gives the corresponding L/D of 1.44 maintaining the purity at its optimal value.

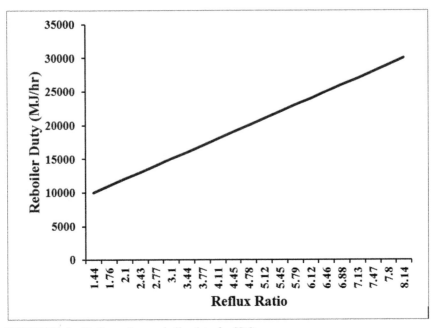

FIGURE 21.9   Reflux ratio vs reboiler duty for SRC.

21.5 CONCLUSION

A steady-state run was simulated on CHEMCAD to separate MCH and toluene mixture using phenol as an extracting agent, using modified UNIQUAC thermodynamic model. The effect of reflux ratio and reboiler duty on the number of plates has been studied. Optimum values for reflux ratio and reboiler duty were determined via rigorous CHEMCAD (6.3.1) simulation runs. Effect of phenol flow rate on MCH purity was also presented in this work. High purity products are obtained in two successive columns and this was showed in the simulation. From the above simulation it was concluded that almost complete recovery of extracting solvent (toluene) can be possible.

KEYWORDS

- **distillation**
- **extracting solvent**
- **CHEMCAD software**
- **optimum flow**
- **reflux ratio**

REFERENCES

1. Linninger, A. *Distillation, Senior Design CHE 396;* 1–30, www.che.utah.edu/~ring/Design I/Articles/distillation design.pdf
2. Olutrotimi, A. D.; Adedayo, O. O. System Identification of Extractive Distillation Process. *Int. J. Eng. Res. Technol.* **2015,** *4,* 145–151.
3. Tiverios, P. G.; Van Brunt, V. Extractive Distillation Solvent Characterization and Shortcut Design Procedure for Methyl Cyclo Heaxane-Toluene Mixtures. *Ind. Eng. Chem. Res.* **2000,** *39,* 1614–1623.
4. Kaushik, V. *Dynamics and Control of Distillation Using Aspen*, Ph.D. Thesis, NIT Raurkela, 2011.

PART IV
Experimental Review

CHAPTER 22

AN EXPERIMENTAL REVIEW OF NON-DESTRUCTIVE TESTING METHODS FOR FRUITS AND VEGETABLES

A. SHARMA*, MD. N. IQBAL, and S. SINGHA

Department of Food Engineering, National Institute of Food Technology, Entrepreneurship and Management (NIFTY), HSIIDC Estate, Kundli, Haryana, India

Corresponding author. E-mail: anshuman.sharma8391@gmail.com

CONTENTS

ABSTRACT

Non-destructive testing (ND) is a method of evaluation of quality of a food product without damaging the material or its suitability for further use. The quality of a food product can be defined as sweetness, absence of physical damage, maturity, etc. Such qualities can be predicted in terms of various physical properties namely, size, shape, color, density, electrical properties, etc. Currently, several ND methods like ultrasonic, infrared (RI) spectrometry, x-ray, magnetic resonance, etc. have been tried at laboratory/commercial level. Each of these methods has its own flaws. To overcome such limitations, two different approaches can be adopted namely, (a) identification of new physical parameters which can correlate with the internal quality more accurately and robustly and (b) development of estimation systems involving more than one physical properties which in combination can predict one or more important internal quality of a fruit or vegetable. Two important ND's for determination of quality of fresh fruits and vegetables have been discussed in this communication. Multi-spectral or hyper spectral image analysis can be used for characterizing size, shape, color, etc. of a fresh fruit/vegetable. Another technique is to use one or more electrical properties like capacitance, resistivity, and inductance to predict internal quality. For either of these techniques, an important aspect is analyzing the data acquired which determines the success of the ND.

22.1 INTRODUCTION

Fresh fruits and vegetables are an important part of a well-balanced regular diet. They contain micronutrients which help in keeping our body healthy. The micronutrients include vitamins like vitamin-A (β-carotene), vitamin-C, vitamin-E, etc. as well as minerals like magnesium, zinc, phosphorus, etc. These components in fruits and vegetables can manage various diseases namely, obesity, cholesterol, blood pressure, heart diseases, and cancer. However, for proper digestibility, palatability, and nutrition the fruits and vegetables need to be consumed in proper physiological state (or maturity) and in safe condition (or without any physical damage). Thus, the quality of fruits and vegetables are necessary to be measured before procurement or use.

The term quality implies the degree of excellence of a product which is often defined by either consumer perception or by end use of the product.

Quality of produce comprises sensory properties (appearance, texture, taste, and aroma), nutritive values, chemical constituents, mechanical properties, and defects. People use all five senses to evaluate quality namely sight, smell, taste, touch, and even hearing. The use of these senses gives information about the appearance, aroma, flavor, hand and mouth feel, and acoustics of the produce. Generally, instrumental measurement is preferred over sensory evaluation of quality in order to reduce variations and to provide more precision and a universally accepted value for all researchers, industries, and consumers. Instruments may be designed to predict the quality categories close to human perceptions and judgments. Some sensors are based on near infrared (NAR), x-ray, magnetic resonance, etc. Various authors have reviewed aspects of these wide varieties of non-destructive testings (ND's) applied to horticulture products.[1-3]

However, each of these methods has their own limitations; some of them are not suitable in commercial environment due to cost or requirement of elaborate facilities whereas some are not robust enough. Hence, there is a need of reliable, scalable, robust, economic, and operationally simple ND's for various horticulture products. Our literature survey has revealed internal quality prediction by two broad ND classes namely, *image processing* and *electrical properties* are very promising to develop better commercially viable sensors for sorting/grading of fruits and vegetables.[3]

The image processing method is one of the non-destructive methods to evaluate the food quality. Image processing technique has an efficient application in the evaluation of food quality and can help in characterizing complex sizes, shape, color, and other properties of food product. The general steps involved in image analysis are as shown in Figure 22.1.

FIGURE 22.1   Five components of image processing technique.

Image acquisition is the first step in the process of image analysis for determining several parameters of food products, which help in evaluating its quality. In this step, the image is first captured and then converted in digital form. The lighting conditions can affect the quality of the captured image. A high quality image can reduce the time and the complexity of the subsequent image if processed by using a proper illumination system and thus can reduce the cost of image processing system.[4] In order to assess composition and quality of food, various attempts and researches have been made in the development of non-destructive non-invasive sensors like charge coupled device (CCD) camera, ultrasound, magnetic resonance imaging (MRI), etc. Most frequently used sensor for determining outer features of fruit during image processing is CCD camera, while internal features can be determined by ultrasound, MRI etc.[5,6] In the image pre-processing step, the captured images, using CCD camera, ultrasound, MRI etc., contain noises and are needed to be processed. The noises present in the images may degrade the image quality, which in turn will not provide correct information about the images. The main purpose of pre-processing is the improvement of image data, which prevent the distortion or enhances some image features in order to make the image fit for further image processing. Generally, the types of image pre-processing techniques used are as pixel pre-processing and local pre-processing. The pixel pre-processing converts the input image into output image in such a way that each output pixel corresponds directly to input pixel having the same coordinates, except the values are modified according to transformation function (hue, saturation, and intensity (HSI) color space transformation using red, green, and blue (RGB) model).[6,7] While, the latter one provides the new values obtained by the averaging of brightness at all the neighborly points, which have similar properties as that of processing points.[8]

In image segmentation step, the image is divided into its constituent objects, which require efficient and skilled work because of the visual information present in the image. This technique can be performed by four approaches namely thresholding-based, region-based, gradient-based, and classification-based segmentation. The most effective approach is thresholding-based technique. The methods used in this step are statistical, fuzzy logic, and neural network.[9]

The image is now successfully segmented into discrete objects, which can be processed and analyzed by measuring characteristics of the object. The external or internal characteristics can be used to describe a segmented object. The characteristics that evaluate the food quality can be grouped as: shape, size, color, and texture.

Size and shape of an object are the basic measurements, needed to determine for quality evaluation. The area, perimeter and length, and width are the three commonly used features for an object size determination. The best convenient method for size is the area and for a pixel-based representation, it is the number of pixels within the area. Keshavarzpour et al.[10] showed that aspect ratio (height and diameter ratio) of apple can be used to classify its size as small (as misshapen) and medium and large (as normal). While, the most commonly used method for shape determination are roundness ratio and ellipsoid ratio. Rashidi and Seyfi[11] showed that roundness ratio and ellipsoid ratio can be used to classify the fruit shape as normal and misshapen fruit.

Color is an influential attribute and a powerful descriptor. Color vision offers a tremendous amount of spatial resolution to quantify the color distribution of ingredients. HSI color space method is mostly used in color evaluation. Iqbal et al.[12] showed that hue mean and hue median can be used to classify the fruit on the basis of maturity. Each pixel of an object is used to examine its color features.

Texture is an attribute representing the spatial arrangement of gray levels of the pixels in a region. Textural property plays an important role in determining the quality of fruit. Arivazhagan et al.[13] proposed an efficient combination of color and texture features for the quality evaluation of fruit. Among the texture analysis methods for food evaluation, most approaches are statistical including pixel-value run length method and the co-occurrence matrix method.

In the final step, the objects are classified into different classes, in which the measured features of a new object are compared with the same features of a known object and determining the particular class to which the new object belong. Among the varieties of approaches taken, statistical, fuzzy logic, and neural network are the main three methods of classification.[14,15] Another method to evaluate the parameters, which helps in determining the food product quality, is electrical method. In this section, the main focus will on the methods namely capacitance, conductance, impedance, and permittivity. Kato[16] proposed an electric method for sorting the fruits on the basis of density, which measures the volume by electric capacity and mass by electronic balance. Impedance is the measure of the opposition that a circuit presents to current, when a voltage is applied. In an alternating current (AC) circuit, impendence comprises of resistance, inductance, and capacitance. Permittivity is the ability of a substance to store electrical energy in an electric field.

This communication deals with methods of non-destructive inspection of internal quality of fruits especially their sweetness and maturity for a large fruit like watermelon. In fruits, sweetness is the major acceptability criteria for the consumers. Sweetness can be measured by total soluble solids (TSS), total reducing sugar, titratable acidity (TA), etc. However, the tests for those properties are destructive in nature which requires damaging of the fruits or vegetables. A brief review of various ND's for fruits and vegetables has been done. In addition to that, two specific methods "image analysis" and "electrical properties" have been studied for suggesting a better methodology for estimating sweetness or maturity for fruits like watermelon.

22.2 MATERIALS AND METHODS

22.2.1 *PHYSICAL ATTRIBUTES AND THEIR DETERMINATION*

22.2.1.1 *FRUITS SHAPE DETECTION*

Theory:

In order to obtain required parameters for fruit shape detection, three perpendicular axes, length (L, mm), major diameter (D, mm), and minor diameter (d, mm) were measured using a digital vernier caliper. An easy technique of judging based on the analysis of geometrical attributes was used to detect shape of fruit. Roundness ratio ($R.R.$) was used to detect oblate, regular, and oblong fruits and is defined as:

$$R.R. = \frac{L}{\sqrt{Dd}}$$

Another parameter, ellipsoid ratio ($E.R.$) was used to detect the round and the elliptical shape of fruit and is defined as:

$$E.R. = \frac{D}{d}$$

Experimental methodology of determination:

The shape of watermelon can be identified by two possible methods namely dimensions of object and image processing. In the former method, the dimensions of watermelon that is major diameter (D) and minor diameter (d) are measured as shown in Figure 22.2. In the latter method, the number of pixels are measured to evaluate the diameter along major and minor axes and finally, to determine the ellipsoid ratio.

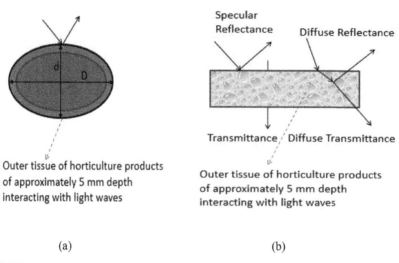

(a)   (b)

FIGURE 22.2 (a) Determination of diameters along major and minor axes; (b) result of incident light on a fruit/vegetable into specular reflectance, diffuse reflectance from features at depths to about 5 mm (body reflectance and interactance), diffuse transmittance or absorbance.

22.2.1.2 DENSITY

Theory:

The property "density" provides the information about the quality of fruits and vegetables. There is always a change in the density of the produce till maturity because of the changes in internal properties of fruits and vegetables like acidity, TSS content, etc. Density helps in measuring hollowness and also in estimating the TSS content, both non-destructively and rapidly. Depending upon the cavity (or density), watermelon is divided into two categories as high-density watermelon, having no cavity, and low-density watermelon, having cavity and small holes around the seeds.

A recent method of electric density sorting is proposed to provide the relation between the hollowness, TSS content, and the density of watermelon using:

$$V.\rho = (V - V_c)\,\rho_c$$

where V—volume of whole fresh sample, cm^3
  ρ—density of whole fresh sample, g/cm^3
  V_c—total volume of void around the seeds, cm^3
  ρ_c—density of pulp, g/cm^3

The measurement of cavity volume of watermelon was done by pouring water into the cavity of fruit in such a way that it penetrates large cavities as well as small holes around the seeds. While, the TSS in watermelon was measured by averaging the result obtained by using refractometer by taking samples of watermelon flesh from five different positions.

Experimental methodology of determination:

The density of watermelon was possible to be determined in different ways non-destructively. For direct density evaluation, a sensitive weighing balance was used to determine the mass of the product. However, the volume was determined by four methods namely volume displacement method, dimensions from image analysis, dimensions using vernier caliper, and capacitance.

In the first method, a beaker is filled with water and the product is dipped in the solvent. Then, the rise in level of solvent was measured, which is equivalent to the volume of the product. The rise in level was measured as:

$$V = A - B$$

where V—rise in level of solvent, equivalent to volume of product

A—level of solvent before dipping the product

B—level of solvent after dipping the product.

In the second method, the volume of the product was evaluated by the dimensions (D, d) obtained from the image analysis (Fig. 22.3). While in the third method, the same dimensions were evaluated by using vernier caliper. This method will provide the possible percentage of error by using image-processing method. In the final method, the capacitance helped in determining the density of the product directly.

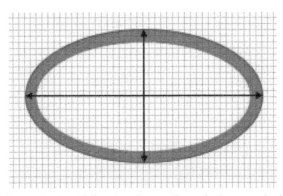

FIGURE 22.3 Determination of diameter along major and minor axes by image analysis.

22.2.1.3   COLOR

Theory:

The color of fruit is considered as the primary critical quality parameter for the acceptance of the product. The first perception that a consumer uses for the acceptance or rejection of the fruits is the color of the product. The determination of color is done by either visually or by using color-measuring instruments. The results obtained by human inspection (visual) vary from observer to observer. Therefore, it is better to use color analysis methods rather than human inspection.

There are different methods to determine the color of the product, but it is recommended to use $L*a*b*$ (Hunter color lab) method, where $L*$ refers to luminance or lightness component ranging from 0 to 100, while $a*$ (from green to red) and $b*$ (from blue to yellow) refers to chromatic components ranging from −120 to +120 and the instrument used to measure the color is colorimeter. Colorimeter provides the detailed description about the fruit through its image and thus, evaluates its quality more precisely, but these instruments have their disadvantages that the measuring surface is to be uniform and small (~2 cm²) and thus, these measurements are not very representative especially in heterogeneous material such as food products.

Recently, the computer vision, a computational technique using a digital camera, is being used to measure the color of the food items by analyzing the pixels of the surface of the food and determining the surface properties and defects. In this method, image-processing software was used to provide measure of the fruit's color by using the digital camera. The measurement of color of any pixel of the image of the object is possible by using three color sensors per pixel. Most commonly used color model is RGB model in which each sensor captures the intensity of a particular light in the spectrum.

Experimental methodology of determination:

The color of watermelon and its variation was possibly evaluated by three methods namely intensity of color, patch color, and shine. The intensity of the product's color was determined by using hunter lab colorimeter. The instrument uses three values L (lightness factor), a, and b (chromatic components) values to determine the color difference between two objects. The L, a, and b values of the standard color are known and the same values of the sample are determined. Then, the difference between the standard values and the determined values, that is, ΔL, Δa, and Δb are evaluated to determine the difference between the color of the object. Similarly, the shine was also determined by using the same hunter lab colorimeter.

While, the patch color is also one of the factors which helps in determining quality of the product. The patch color is the ratio of the area of other colors (yellow in case of watermelon) to the color of the product (green in case of watermelon), which was evaluated with the help of ImageJ software (Fig. 22.4).

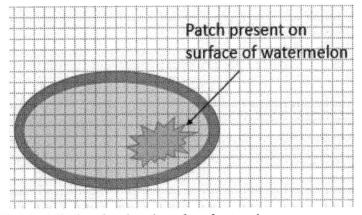

FIGURE 22.4   Indication of patch on the surface of watermelon.

ImageJ software converts the number of pixels into unit distances (cm, mm, etc.) and the dimensions are represented in those units. Using the software, the projected area of the fruit and the patch can be determined. The ratio of the area is later correlated with the internal qualities.

22.2.1.4   ELECTRICAL METHODS

Theory:

Capacitance and conductance are two electrical properties that can be used to determine internal quality of horticultural products. The term capacitance means the ability of a body to store an electrical charge at a given voltage and can be measured only with AC source. Higher the capacitance, higher the capacity to store the electric charge. Electrical conductance is the inverse of electrical resistivity and can be measured as the ease in the passage of electric current through the conductor. Capacitance can be measured with AC source.

The model to measure capacitance comprises of two parallel conductive plates, consists of aluminum material because of its property of being not easily ionizable, voltage input, and two impedances (mainly capacitance) as reference and sensing. The product is placed between the plates of sensing impedance. Soluble solid content, TSS is the main quality of fruit detected by capacitance. With increase in capacitance of the product, the ripening of the fruit increases with increase in the soluble solid content. During the ripening of fruit, not only the starch present gets converted into sugar but also the firmness of product changes. Increase of TSS content is an important trait of hydrolysis of starch into soluble solids such as glucose, sucrose, and fructose.

Experimental methodology of determination:

Capacitance of watermelon was determined using the experimental set up given in Figure 22.5 and the capacitance individually or in combination with other properties in determining the density and other internal quality of the product directly. The model used to measure the density by using the capacitance consists of two electrodes as shown in figure below. The capacitance depends upon the charges on the electrodes as well as the dielectric material between them. Generally, the dielectric material is air and if the dielectric material changes, the capacitance also changes. In this study, the dielectric material is replaced by watermelon.

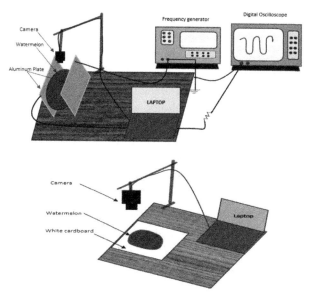

FIGURE 22.5   Experimental setup of non-destructive testing of watermelon.

22.3   RESULTS AND DISCUSSION

NDT of larger fruits with thicker skin is a challenging task, especially for products like watermelon. Many authors have reported non-destructive determination of qualities like TSS, firmness, internal defect, and maturity of watermelons. For such estimations, the technologies used are acoustic technology, dynamic technology, electrical and magnetic technology, x-ray and computed tomography, and NIR spectroscopy, etc. Some of them can be used for online measurement or sorting of melons. However, they all have their own limitations and they are much sophisticated to use in small scale or to use in procurement sites to grade the watermelons. In this study, an attempt has been made to identify the major physical variables that can predict the sweetness of watermelon which is the major criteria to grade the fruit for fresh consumption or for processing.

22.3.1   *FORMULATION OF THE OBJECTIVE SWEETNESS INDEX*

Sweetness is a subjective quality parameter of fruits and its measurement is usually influenced by personal liking/taste sensitivity as well as other intrinsic or extrinsic factors. For example, color, an intrinsic factor of fruit influences perception of sweetness of fruits (color influences sensory perception and liking of orange juice-ref). External factors like temperature too alter the perception of sweetness. In this study, internal qualities like TSS, reducing sugar, TA, and pH of the watermelon juice were measured and compared with subjective sweetness. To determine subjective sweetness, a group of 50 people scored sweetness of the samples on a 0–9 point scale. The average of the 50 scores for each sample was considered as the sweetness score of the sample. Often TSS or TSS/TA is quoted as two internal quality factors that correlate well with maturity (and sweetness) of various fruits. Watermelons of a particular variety but of different ages were taken in this study. Table 22.1 shows the TSS, acidity as well as subjective sweetness score of various samples of the fruit. However, none of these internal qualities were correlating with the sweetness scores, and the r^2 values were lesser than 0.5.

The acidity and reducing sugar content of watermelon depends upon the maturity stage. Usually with increase in maturity of the fruit, TSS is expected to increase whereas the acidity to decrease. Based on which watermelon can be classified into mature, ripe, and over-ripe categories.

Some authors have observed a strong correlation of sweetness with reducing sugar content (RS) (ref). In this study, a so-called objective sweetness index (OSI = TSS × RS/TA) was found to have an excellent correlation with subjective perception of sweetness (Fig. 22.6). The relation between OSI and subjective sweetness index was found to be non-linear as: $y = 4.440 \ln(x) + 12.81$; but the r^2 value was high as 0.958.

TABLE 22.1 Variation of Subjective Sweetness with Respect to Titratable Acidity and Total Soluble Solids.

S. No.	Total soluble solids (TSS) Bx%	Titratable acidity (TA)	Average score of sweetness
1	9.25	5.6	25.6
2	7.45	4.2	19.2
3	7.97	5.9	12.8
4	6.31	8.7	9.6
5	5.96	7.0	16
6	8.26	6.7	25.6
7	7.01	7.3	19.2
8	7.82	7.2	19.2
9	8.08	8.4	25.6
10	6.47	4.2	44.8
11	6.81	6.4	22.4
12	6.65	5.9	25.6
13	7.16	6.2	22.4
14	7.06	4.9	25.6
15	6.92	6.9	19.2
16	7.51	6.2	19.2

This ratio is a better predictor of sweetness because TSS represents not only different kinds of reducing (fructose, glucose, etc.) and non-reducing (oligosaccharides from hydrolyzed polysaccharides) sugars but also other soluble components which individually may not be sweet in taste. In addition, sweetness suppressing function of titratable acidity might be dependent on both TSS and reducing sugar. In subsequent studies, OSI has been used to correlate with physical variables (e.g., size, shape, skin color, etc.) to reduce the effect of the variability of subjective evaluation of sweetness on the model usable for NDT of sweetness.

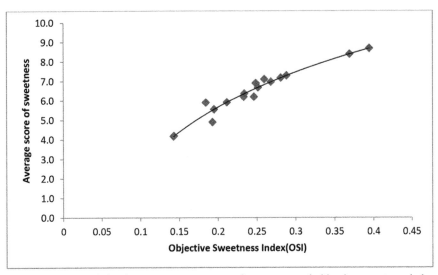

FIGURE 22.6   Relation between average score of sweetness and objective sweetness index (OSI).

22.3.2   ANALYSIS OF SKIN COLOR

In typical image analysis of horticulture products, visible light or IR waves are used. When such electromagnetic waves fall on the product, some percentage (about 4%) of light gets reflected from the surface (known as specular reflectance) and the remaining portion gets transmitted into the fruit. A portion of the transmitted light, so-called diffused light, reflects back from the depth of around 5 mm. The rind part of a watermelon is very thick compared to fruits with thin skin like apple (can be as high as greater than 24 mm) (Fig. 22.7). Hence, the IR-based reflectance measurement fails to give any information about the internal quality of watermelons.

Therefore, two different strategies of skin-color-based evaluation of maturity and in turn sweetness of watermelon were adopted in our study. Complete skin color was analyzed with a colorimeter, and also from image capture of the fruit, the ratio of yellow patch and total projected area of the fruit was estimated.

The *L*, *a*, and *b* values of different watermelon samples were determined by using hunter color lab colorimeter. In this method, the scale to measure lightness (*L*) varies from 100 (perfect white) to 0 (black). The chromaticity factors "a" reads positive as redness, zero as gray, and negative as greenness, while "b" shows positive as yellowness, zero as gray, and negative as blueness. The standard black and white tiles were used to calibrate colorimeter and the values displayed on the screen of colorimeter were matched. The nose cone was in contact with the surface of fruit to prevent the leakage of light emitted by colorimeter. The values of L, a, and b obtained from four different places, that is, near the apex and stem regions of mango, were averaged and then used for further calculation.[17] However, the skin color indices of the samples showed very poor correlation with the OSD.

Often the maturity of watermelon is predicted in terms of the area of the yellow patch. When the fruit stays for longer time in the field, the yellow patch or the area away from the sun exposure widens in size. Since watermelon is a non-climacteric fruit, a prematurely harvested watermelon is expected to show smaller yellow patch: total projected area ratio and lesser sweetness. In this work, the ratio of the area of patch surface to the area of actual surface was obtained by image analysis using commercial software called ImageJ.

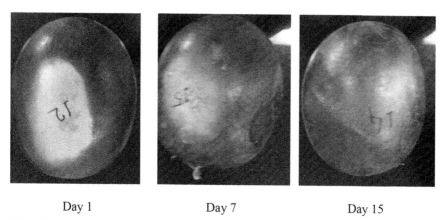

Day 1   Day 7   Day 15

FIGURE 22.7 Changes in the patch color during storage of watermelon.

Our study showed that the difference of patch intensity is highly influenced by storage condition and time. Showalter has also reported such changes in the pigmentation of the watermelon skin. Hence, the ratio is not a reliable marker of sweetness.

22.3.3   IMAGE PROCESSING FOR SHAPE AND SIZE CHARACTERIZATION

The images captured were analyzed by ImageJ software to characterize the shape and size of watermelon. The shape of watermelon was defined by using the ratio of major and minor diameter, which was further determined by two methods, that is, using image analysis and calculated dimensions. In the former method, the measurements were taken after the analysis of image in the software to determine the ratio of major and minor diameter and thus to evaluate the shape by their ratio. While, the dimensions obtained by direct measurement using vernier caliper, were used to determine the ratio of major and minor diameters.

The size or volume of watermelon was also analyzed by the same software and evaluated by two methods as follows: ratio of actual volume and projected volume. In this method, the actual volume was calculated by using volume displacement method and volume was calculated by image analysis. However, OSI was not correlating individually with neither D/d nor the actual and calculated volume ratio (Table 22.2).

TABLE 22.2   Variation in OSI with Respect to Shape and Size.

S. No.	RS*Bx/A	D/d	V_{act}/V_{cal}
1	0.20	1.60	0.51
2	0.14	1.67	0.70
3	0.21	1.78	0.68
4	0.39	1.31	0.46
5	0.27	1.49	0.49
6	0.25	1.59	0.50
7	0.29	1.59	0.54
8	0.28	1.43	0.67
9	0.37	1.56	0.63
10	0.14	1.57	0.75
11	0.23	1.43	0.66
12	0.18	1.70	7.03
13	0.23	1.54	0.53
14	0.19	1.68	0.74
15	0.25	1.51	0.80
16	0.25	1.60	0.72

22.3.4 ELECTRICAL METHOD

The electrical properties can be used to determine the internal quality of watermelon like density, TSS, firmness, maturity, etc. An impedance analyzer was used to determine the density and TSS of the product.[18] The analyzer used the dielectric properties in a frequency range of 10–1.8 GHz. The best correlation coefficients found between the dielectric constant, loss factor, and TSS were 0.60 and 0.72. Also, it was noted that the individual correlations between dielectric constant, loss factor, and TSS were low, but the indices between ε'/SSC and ε''/SSC were high.

In this study, the density which was the indication of quality of watermelon was correlated with the electrical property, that is, capacitance. In order to determine the capacitance, the watermelon was placed between two electrodes and the current was allowed to flow through the electrodes and finally, the density was correlated with capacitance per unit area (nF/A) (Table 22.3).

TABLE 22.3 Variation in OSI with Respect to Density and Capacitance.

S. No.	RS*Bx/A	Density	nF/A
1	0.20	0.91	0.000436
2	0.14	0.88	0.000438
3	0.21	0.88	0.000388
4	0.39	0.92	0.000471
5	0.27	0.96	0.000378
6	0.25	1.09	0.000342
7	0.29	0.88	0.000295
8	0.28	0.78	0.000262
9	0.37	0.98	0.000356
10	0.14	0.92	0.000570
11	0.23	0.93	0.000513
12	0.18	0.09	0.000303
13	0.23	0.88	0.000383
14	0.19	0.88	0.000662
15	0.25	0.87	0.000519
16	0.25	0.85	0.000747

The density of watermelon,[19] obtained after processing the experimental data, was found to be distributed between 0.7 and 1.1 g/cc. In this section, the point we try to discuss is the relationship between the degree of hollowness (voids), density, reducing sugars, TSS contents, acidity, and the sweetness.

The relationship between the density and the degree of hollowness can be shown in Figure 22.8. The figure showed that the percentage of air voids varies as the fruit reaches to its maturity and gains sweetness. As the number of voids decreases, the TSS and the density of the product increases as predicted from the figure. Thus, higher the cavities and small holes (or voids), smaller the density.

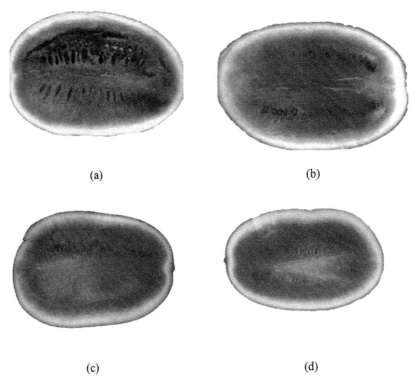

(a)   (b)

(c)   (d)

FIGURE 22.8   Relationship between density, degree of hollowness, and total soluble solids.

Based on the relationship between the density and soluble solid content, the watermelon can be divided into two classes as higher density and lower density watermelons. The watermelon with lower density have higher cavities, while the watermelon with high density neither have cavities nor small holes. Finally, it was predicted that there was an optimum range of density and the quality of watermelon was classified as mature, ripe, and over-ripe on the basis of density.

22.4 CONCLUSION

This study provides the information about the parameters, density, TSS content, and acidity, which affect the quality of watermelon. The degree of hollowness was found to be related to density as low-density watermelon, which has cavities and small holes around the seeds of watermelon, and the high-density watermelon, which has no or less cavities. The sweetness was found to be correlated with the reducing sugar, TSS contents, and acidity in a non-linear way which was considered as a OSI. The OSI was directly related to the TSS content but was indirectly related to the acidity. Thus, it is not possible to predict the quality (or sweetness) of watermelon on the basis of a single parameter but a composite function is probably required.

KEYWORDS

- non-destructive testing methods
- micronutrients
- fruits and vegetables
- image processing techniques
- patch color

REFERENCES

1. Abbott, J. A. Quality Measurement of Fruits and Vegetables. *J. Postharvest Biol. Technol.* **1999,** *15,* 207–225.
2. Nicolai, B. M.; Beullens, K.; Bobelyn, E.; Peirs, A.; Sacy, W.; Theron, K. I.; Lammertyn, Non-Destructive Measurement of Fruits and Vegetables Quality by Means of NIR Spectroscopy: A Review. *J. Postharvest Biol. Technol.* **2005,** *46,* 99–118.
3. Sun, D. W.; Du, C. J. Recent Development in the Application of Image Processing Techniques for Food Quality Evaluation. *Trends Food Sci. Technol.* **2004,** *15,* 230–249.
4. Gunasekaran, S. Computer Vision Technology for Food Quality Assurance. *Trends Food Sci. Technol.* **1996,** *7,* 245–256.

5. Rigney, M. P.; Brusewitz, G. H.; Kranzler, G. A. Asparagus Defect Inspection with Machine Vision. *Trans. ASAE.* **1992,** *35,* 1873–1878.

6. Li, Q. Z.; Wang, M. H.; Gu, W. K. Computer Vision Based System for Apple Surface Defect Detection. *Comput. Electron. Agric.* **2002,** *36* (2–3), 215–223.

7. Tao, Y.; Heinemann, P. H.; Vargheses, Z.; Morrow, C. T.; Sommer, H. J. III Machine Vision for Colour Inspection of Potatoes and Apples. *Trans. ASAE.* **1995,** *38,* 1555–1561.

8. Utku, H.; Koksel, H. Use of Statistical Filters in the Classification of Wheat by Image Analysis. *J. Food Eng.* **1998,** *36,* 385–394.

9. Zion, B.; Chen, P.; McCarthy, M. J. Detection of Bruises in Magnetic Resonance Images of Apples. *Comput. Electron. Agric.* **1995,** *13,* 289–299.

10. Tabatabaeefar, A.; Rajabipour, A. Modeling the Mass of Apples by Geometrical Attributes. *Sci. Hort.* **2005,** *105* (3), 373–382.

11. Rashidi, M.; Seyfi, K. Classification of Fruit Shape in Cantaloupe Using the Analysis of Geometrical Attributes. *J. Agric. Sci.* **2007,** *3* (6), 735–740.

12. Iqbal, S. Md.; Gopal, A.; Sankarnarayanan, P. E.; Nair, A. B. Classification of Selected Citrus Fruits Based on Colour Using Machine Vision System. *Int. J. Food Properties.* **2016,** *19,* 272–288.

13. Arivazhagan, S.; Shebiah, R. N.; Nidhyanandhan, S. S.; Ganesan, L. Fruit Recognition Using Colour and Texture Features. *J. Emerg. Trends Comput. Infor. Sci.* **2010,** *1* (2), 90–94.

14. Park, B.; Chen, Y. R.; Nguyen, M.; Hwang, H. Characterizing Multispectral Images of Tumorous, Bruised, Skin-Torn and Wholesome Poultry Carcasses. *Trans. ASAE.* **1996,** *39,* 1933–1941.

15. Hu, B. G.; Gosein, R. G.; Cao, L. X. de Silva, C. W. Application of Fuzzy Classification Technique in Computer Grading of Fish Products. *IEEE Trans. Fuzzy Syst.* **1998,** *6,* 144–152.

16. Kato, K. Electrical Density Sorting and Estimation of Soluble Solids Contents of Watermelon. *J. Agric. Eng.* **1997,** *67,* 161–170.

17. Jha, S. N. Chopra, S.; Kingsly, A. R. P. Modelling of Colour Values for Nondestructive Evaluation of Maturity of Mango. *J. Food Eng.* **2007,** *78,* 22–26.

18. Nelson, S. O. Dielectric Properties of Agricultural Products and Some Application. *J. Agric. Eng.* **2007,** *54,* 104–112.

19. Kato, K. Non-Destructive Measurement and Quality Evaluation of Agricultural Products Constituent, Structure and Temperature. *Syst. Control Inform.* **1996,** *40* (2), 63–68.

CHAPTER 23

NOVEL TECHNOLOGY FOR ESSENTIAL FATTY ACIDS: AN EXPERIMENTAL REVIEW

S. P. NAYAK* and S. SINGHA

National Institute of Food Technology Entrepreneurship and Management, Plot No. 97, Sector 56, HSIIDC Industrial Estate, Kundli, Sonepat 131028, Haryana, India

Corresponding author. E-mail: siba.prasad27@gmail.com

CONTENTS

ABSTRACT

Omega-3 and omega-6 fatty acids are known as essential fatty acids (EFAs) due to the inability of human body to synthesize them. Hence, their supply to the body has to be through diets. There is a gamut of literature available on health benefits of EFAs namely, mental stress management, lowering of serum triglyceride levels in blood, etc. Some of the major sources of these fatty acids are fish oil, flaxseed oil, and algae oil. In commercial process, these two EFAs can be served as concentrated nutraceutical (in powder or liquid form) or can be incorporated in other food systems. However, in both cases, oxidation of the acids is a prime challenge. In addition to that, there are other challenges like off-flavor of the oil, ensuring appropriate level of the acids in each serving plus their stability in the food matrix. To meet these challenges, a study has been designed to develop an optimal process to produce microencapsulates of EFAs. The method essentially involves encapsulation via homogenization followed by a freeze/spray drying. In this study, appropriate selection of encapsulation matrix as well as optimization of the process conditions has been done.

23.1 INTRODUCTION

Omega-3 long chain polyunsaturated fatty acid (PUFA) like eicosapen-taenoic acid (EPA) and docosahexaenoic acid (DHA) are well known for their role in building healthy body and developing body immunity to defend against diseases. Health benefits of omega-3 fatty acids include reducing blood pressure, reducing heart disease (coronary heart disease (CHD)), reducing stiffness and joint pain (rheumatoid arthritis), curbing depression, improving brain and visual in infants, and reducing asthma. These essential fatty acids (EFAs) cannot be synthesized by human body thus need to be taken from outside in form of capsule, soft gels, and in food diet. Major sources of these EFAs include oil from marine fishes like tuna, mackerel, menhaden, anchovy, cod, herring, sardine, salmon, capelin, rainbow trout, etc., flaxseed, and algae. According to *European Union* (EU), daily intake of 1 g of these EFAs is required to maintain healthy and diseases-free body. The US Food and Drug Administration (FDA), in year 2002 certified a health claim of these EFAs from fish oil of "consumption of omega-3 fatty acids may reduce the risk of coronary heart disease" by approving it with a guide-line, a product must contain at least 600 mg of these EFAs and must not exceed 2000 mg per day. It was found out that at least two fish meals per week are required to achieve the require amount of EFAs for our healthy body. However, despite having certified health benefits, consumption level of these EFAs to give health benefit is quiet below the desire level. The reasons may be due unavailability of fresh fish, fishy aroma, and culture. Thus, extracting oil from fish and incorporating into food could solve the problems. However, these EFAs are highly susceptible to oxidation, as they are highly unsaturated with five or six double bonds, and fishy aroma restricted its incorporation into food matrix. Solution to oxidation could be hydrogenation, but reducing its degree of unsaturation ends up with loss of beneficial effects of these EFAs. To overcome above two challenges, microencapsulation of fish oil is essentials.

Microencapsulation of fish oil can be done by various methods such as freeze drying, spray drying, coacervation, etc. Microencapsulation process primarily comprises of two steps, first emulsion formation and second, conversion of emulsion into powder, beads, particles, or pellets, depending upon methods employed for encapsulation. Emulsion formation with oil droplets' size ranging into micro range requires suitable selections of wall materials with appropriate quantities, emulsifier and its quantity if wall material dose not poses surface-active property, and emulsifying methods. Choice of suitable wall material is the key to successful microencapsulation of fish oil. Wall material used for encapsulating purposes should pose high water-soluble, good emulsifying property, low viscosity, good drying property, and good barrier to oxygen transfer between surrounding and oil droplet when dried. Wall materials commonly used in microencapsulation of fish oils are proteins, such as, gelatin, sodium caseinate, soy proteins, and whey proteins, carbohydrates, that is, modified starches and emulsifying starches, small-molecule carbohydrates, such as, maltodextrin, sucrose, glucose, and lactose, and gums, that is, gum arabic. Pectin, gelatin, and maltodextrin as wall materials provide advantages of good encapsulation efficiency with good oxidation stability. Pectins are hetero-polysaccharides with molar mass in the range of 10^4–2×10^5 g/mol. Alpha (1–4) D-galacturonic acid units constitute 65% weight of pectins. Pectins are soluble in water and form solution in the concentration range of 6–12% depending upon the types of pectin. Pectin solutions are less viscous and usually having gelation properties. Pectins can be obtained from apples, pears, plums, gooseberries, guavas, quince, and other citrus fruits, and also in lesser extent can be obtained from grapes, cherries, and strawberries. Gelatin consists of both single- and multi-stranded polypeptides. Each polypeptide consists of 300–400 amino acids. Gelatins are amphiphilic and in aqueous solution, depending upon pH, it behaves as polyelectrolytes. Gelatins have the property of getting solidified upon cooling and melted upon heating. Gelatin can be obtained from fish skin, pork skin, and bones of horses, cattle, and pork.

Emulsion formation can be carried out by different methods such as high-speed mixer, colloid mill, high-pressure valve homogenizer, ultrasonic homogenizer, microfluidization, membrane homogenizer, and microchannel homogenizer. Microfluidization from aforementioned methods poses distinguished advantages of producing droplets of size <0.1 μm and uniform droplet-size distributions.[1] Very small particle size can be generated by allowing coarse emulsion to pass through the microfluidizer a numbers of times or by allowing the pump pressure to increase such that emulsion fluid can be accelerated inside the channel to a very high velocity.[2] Above all, with industrial microfluidizer, an operating pressure of 275 MPa and throughput of 12,000 lh^{-1} can be achieved.[1] Thus, the role of microfluidizer in microencapsulation is tremendous as it can produce droplets with size ranging in micro to nano ranges with mono-disperse type of particle size distributions. Most commonly employed methods to convert fish oil emulsion into powder were spray drying, freeze drying, and coacervation. Apart from aforementioned technologies, melt injection, extrusion, submerged co-extrusion, and cyclodextrins are generally practiced to produce water-soluble microencapsulates.[3] Freeze drying method poses additional advantage of highly stable microencapsulates for which it is usually considered as an ideal method to prepare microencapsulates. In this study, two cod liver oil microencapsulates were prepared by using two different wall material compositions, first wall material being consisting of pectin and maltodextrin and second wall material being composed of pectin, gelatin, and maltodextrin followed by its comparison study with others studies. Emulsion preparation was carried out by microfluidizer and conversion of emulsion from liquid to dried powder was done by freeze drying. An experimental verification of the most promising review for EFA is reported in Table 23.1.

23.2 THEORETICAL BACKGROUND

23.2.1 *METHODS OF EMULSION PREPARATION*

23.2.1.1 *COLLOID MILLS*

In food industry, colloid mills have application of secondary homogenization, that is, homogenization of medium and highly viscous liquids like meat or fish paste and peanut butter.[4] Colloid mills usually consist of two discs, a rotor (rotating disc) and a stator (static disc). Colloid mills are used to homogenize coarse emulsion, rather than to homogenize separate phases, that is, oil and water phases as it is more efficient in homogenizing coarse emulsion rather than doing primary emulsion. To prepare coarse emulsion from oil phases and water phases, high-speed mixers are usually used. Droplet-size reduction is usually achieved by rapid rotation of rotor inside the stator as high shear stress is produced between the gap of rotor and stator. Rotation of rotor also generates centrifugal force which moves fluid toward the peripheral zone, where it is extracted for further use. Variable shear stress can be achieved by three ways: by varying the gap thickness between stator and rotor (50–1000 μm), by varying the speed of rotor, and by using roughened rotor or interlocking teeth.[5] Two mechanisms are generally followed in colloid mills to disrupt droplets: first, laminar shear flow if the surface is smooth and second, turbulent shear flow if the surface is rough.[6] Droplet disruption can be enhanced by allowing liquid to spend more time in the mill which in turn can be achieved by reducing the coarse emulsion flow rate (4–20,000 lh⁻¹) or by passing the emulsion through mills for multiple times.[1] Food manufacturers must therefore put more emphasis on rotation speed, type of stator, and throughput that yields the best result. There should be provision to set cooling device to offset temperature increase due to viscous dissipation losses. Usually droplet size between 1 and 5 μm can be achieved by using colloid mills[1].

23.2.1.2 HIGH-PRESSURE VALVE HOMOGENIZERS (HPVH)

In food industry, to produce fine emulsions, high-pressure valve homogenizers (HPVH) are more commonly used and similar to colloid mills, HPVH are commonly employed for secondary homogenization.[7] Similar aforementioned method is used to prepare coarse emulsion, and is followed by feeding it directly as an input to HPVH to prepare fine emulsion. The homogenizer pump pulls coarse liquid into a chamber on its backstroke and then forces the liquid to pass through a narrow valve on its forward stroke. The flowing coarse emulsion in the valve experiences disruptive forces which cause the larger droplets to be disintegrated into smaller droplets. The flow regime depends on the design of homogenization nozzle, material being homogenized, and size of homogenizer.[7] Different types of application-specific nozzles have been come into use to increase the efficiency of droplet disruption. Most homogenizers are equipped with spring loaded standard valves in order to vary the gap through which emulsion passes. Decreasing gap size causes greater degree of droplet disruption but increases manufacturing costs. Efficiency of droplet disruption also depends on dimensions of homogenizers. In some commercial homogenizers, a two-stage homogenization process is achieved by forcing the emulsion through two consecutive valves. First valve causes the breaking of droplets due to high operating pressure while the second valve operating at lower pressure disrupts "flocs" if formed after first stage. These types of homogenizers are mainly used in dairy industry. If oil and liquid are mixed prior to homogenization, it is possible to create submicron size emulsion. If very fine emulsion droplets are to be produced, liquid is to be passed through homogenizer a number of times. In the presence of high emulsifier amounts, fine droplets of about 0.1 μm range can be produced by HPVH[1]. Temperature rise is fairly small, but can be of appreciable range under very high pressure. Hence, the use of water-jacketed homogenization chamber is recommended to prevent heat surge. Nowadays, industrial HPVH are available with throughputs of 100–20,000 lh^{-1} and pressure of 3–2 MPa[1].

23.2.1.3   ULTRASONIC HOMOGENIZERS

This type of homogenizers uses sound waves having frequency more than 20 kHz that generate high pressure and shockwaves resulting from cavitation and turbulent effects to reduce droplet size. Piezoelectric transducers and liquid jet generators are more commonly used for generating high intensity ultrasonic waves in food industry. Small volumes of emulsion can be best prepared by piezoelectric transducers (few 100 cm^3)[1]. A piezoelectric transducer consists of piezoelectric crystal provided inside a metal casing having tapered ending. Ultrasonic waves are generated when high intensity electric wave is given to transducer causing the piezoelectric crystal to rapidly oscillate at more than 20 kHz. When tapered ended transducer is kept inside a liquid container, it produces shock waves at its vicinity responsible for disrupting droplets; therefore, thorough mixing of liquid inside the container is necessary as a basis to obtain uniform droplet-size distribution. Ultrasonic homogenization is used for preparing emulsions in batches, although in-line flow through versions has also been developed. In food industry, ultrasonic jet homogenizers are mainly used for food emulsions preparation. When a fluid stream is made to strike on a sharp-edged blade, it causes the blade to rapidly vibrate, thus generating an intense ultrasonic field which is able to break up any droplets in its immediate vicinity. It can be used for continuous production of emulsions and is more efficient than HPVH for producing same droplet size. A flow rate between 1 and 500,000 lh^{-1} can be achieved with liquid jet type of ultrasonic homogenizer[1]. Most commonly used frequency to produce emulsion is between 20 and 50 kHz, as higher frequency decreases homogenization efficiency[1]. Droplet size produced depends on duration of application of ultrasonic waves and also on ultrasonic waves' intensity.

23.2.1.4   MICROFLUIDIZERS

Microfluidizers are the microchannel-based high-pressure homogenizer. Microfluidizer has three important components, that is, motor, intensifier pump (single acting), and interaction chamber. Interaction chamber is the heart of microfluidizer. Electric motor runs intensifier pump. Intensifier pump during its pressure stroke applies desired pressure to the liquid stream to drive it to interaction chamber. Interaction chamber has a fixed geometry, that is, Z type or Y type microchannels, which is primarily responsible for liquid streams' accelerated velocity. Velocity of 500 m/s can be achieved inside the interaction chamber. High velocity liquid streams collide with each other and also on microchannel surfaces which results in high shear and impact forces. Above forces bring out desired changes in the liquid streams. As pressure strokes end, reverse stroke begins which draws in fresh liquid streams. Once again, the above process is repeated. Same liquid streams can be passed through microfluidizer repeatedly to obtain desired characteristics in the product streams.

Apart from above-mentioned methods for emulsion preparation, other methods are also available, such as, high-speed mixers, membrane, and microchannel homogenizers.

23.2.2   METHODS FOR LIPID/OIL FILLED HYDROGELS

23.2.2.1   FREEZE DRYING

Freeze drying is an ideal technology for drying heat sensitive ingredients, as drying is usually carried out in low temperature. Freeze drying generally consists of three important components: drying chamber, vacuum pump, and condenser (cold trap). Drying chamber allows heat transfer to food materials mainly by conduction and radiation mode. Temperature of cold trap is usually maintained at lower temperature relative to drying chamber.

Vacuum pump usually maintains a constant pressure below the triple point pressure of water, and in most of the cases, pressure is maintained between 100 and 50 Pa. Freeze drying of any liquid material is generally carried out in three steps, first being preconcentrating, second being freezing, and third being drying, consisting of both primary and secondary drying. Each operation like freezing and drying is usually carried out below glass transition temperature. Preconcentrating is done to decrease the moisture content and in turn it elevates the glass transition temperature, that is, if initially glass transition temperature was −30, then it will increase to −20 or like that. Freezing is the most important step of freeze drying. In this process, the material to be freeze dried is kept below its triple point. The sublimation will occur in its succeeding steps. This is the most critical process as the rate of cooling influences the size of ice crystal forms; thus if improperly done, it can lead to spoilage of the product. In primary drying, unbound water was removed by sublimation process and in secondary drying bound water was removed. Secondary drying is generally carried out at higher temperature and higher vacuum pressure, as it is responsible for removing bound water. By freeze drying, more than 95% moisture can be removed. Since every steps of freeze drying is carried out below its glass transition temperature, products' morphology and nutritional value hardly get affected.

23.2.2.2 SPRAY DRYING

Spray drying is the most commonly used method to encapsulate desired material. Spray drying of desired material is achieved by three steps, forming emulsion, atomization of emulsion, and finally spraying of emulsion into a hot chamber. Size of droplet forms generally depends up on liquid properties, which are, viscosity and surface tension, pressure drop in nozzle, and velocity at which liquid sprayed. Droplets temperature should not exceed 100°C. In most cases, the inlet temperature of hot air is about in the range of 150–220°C, outlet temperature in the range of 50–80°C, residence time in the range of 5–100 s, and particle size in the range of 10–300 μm. Nozzles generally used in spray drying are rotary disk, single fluid high-pressure swirl nozzle, two fluid nozzle, and ultrasonic nozzle. In most cases, concurrent flow of hot air and liquid droplets is practiced and it was observed that dried droplets obtained in concurrent flow are more porous compared to counter current flow.

Apart from above-mentioned methods for oil/lipid filled hydrogels, other methods are also available, such as, melt injection, melt extrusion, co-extrusion, coacervation, inclusion complexation liposome entrapment, etc.

23.3 MATERIALS AND METHODS

23.3.1 MATERIALS

Gelatin (SRLchem, Sisco Research Laboratories Pvt. Ltd.), pectin (Titan Biotech Ltd), and maltodextrin (Himedia) were collected from NIFTEM Dairy Engineering Laboratory. Cod liver oil (Seacod, Sanofi India Ltd.) was purchased from Novelties Corner, CP, New Delhi. Hexane (SRLchem, Sisco Research Laboratories Pvt. Ltd.) and iso-propanol for determining surface oil content and total oil content were collected from NIFTEM Chemistry lab.

23.3.2 EMULSION PREPARATION

Two different fish oil emulsions having 30% total solid at 10% w/w oil loading were prepared. Two different emulsions were sample 1, 4% pectin, 16% maltodextrin, and 10% oil, and sample 2, 2% gelatin, 2% pectin, 16% maltodextrin, and 10% oil. All the measurements were carried out on w/w basis.

Sample 1: Pectin was added to distilled water and stirred at 70°C until it was completely dissolved. Then maltodextrin was added to above prepared solution and subsequently was followed by decreasing solution temperature to 60°C. Fish oil was heated to 60°C, and was followed by its mixing with above solution by a rotor stator type mixer (IKA T18 digital ULTRA TURRAX) at 25,000 rpm for 2 min to obtain coarse emulsion. Coarse emulsion was then passed through microfluidizer (Microfluidics M-110P) at 30 MPa for three pass to obtain final emulsion. Sample 2: It was prepared in the same way as that of sample 1 with a slight change of adding gelatin into the solution prior to the addition of maltodextrin.

23.3.3 FREEZE DRYING OF EMULSION

Emulsion was allowed to freeze at −80°C for 24 h in a deep freezer. Freezed emulsion was then freeze dried at −45°C for 70 h by a lab grade freeze dryer (Mini Lyodel, Delvac)

23.3.4 CHARACTERIZATION OF ENCAPSULATED EFA

Moisture content:

Moisture contents of freeze-dried samples were determined by using MOC63u UniBloc moisture analyzer.

Water activity:

Water activities of freeze-dried samples were determined by using Aqua lab dew point water activity meter 4TE.

Color of freeze-dried sample:

Colors of freeze-dried samples were determined by Chroma Meter CR-400, Konica Minolta.

Density:

Bulk density: Three grams of each sample was taken and was filled separately into a 25 mL measuring cylinder via funnel. Then volume was determined by measuring the level to which sample was filled, followed by determination of bulk density by using formula:

$$\text{Density} = \frac{Mass}{Volume} \qquad (23.1)$$

Tapped density: Above samples were taken and tapped several times till no changes in the level of samples were observed. In the same way as that of bulk density, tapped density was calculated.

True density: True density was calculated by toluene displacement method. First, 2 g of encapsulate was taken and was allowed to get fully saturated with toluene by adding it to 5 mL of toluene. Fully saturated encapsulate was dried for a few minutes to remove surface toluene. Then, encapsulate was added to 5 mL of toluene, followed by measuring the increase in height of toluene level in measuring cylinder. Then true density was determined by calculating the ratio of weight of the sample taken and volume of toluene displaced.

23.3.5 ENCAPSULATION EFFICIENCY

Surface oil loading:

Surface oil of freeze-dried sample was determined by using method followed by Ref. 8. Method in brief: First, 30 mL of hexane and 1 g of freeze-dried sample were taken. Freeze-dried sample was allowed to disperse in hexane with immediate action of vigorous shaking for 30 s. Solvent was filtered out by filter paper into a 100 mL beaker. Solvent in beaker was allowed to evaporate overnight by placing beaker with solvent in a fume hood. Remaining solvent was removed by heating beaker to 105°C for 30 min. Surface oil of freeze-dried sample was then calculated gravimetrically.

Total oil content:

Total oil content was determined by using the method followed by Ref. 8 with slight modification. Method in brief: Two grams of freeze-dried sample was added into 8 mL of distilled water and then mixed at 300 rpm for 2 min. Hexane/iso-propanol (3:1 V/V) solution was prepared. The resulting solution after mixing and 25 mL of hexane/iso-propanol solution was then added together, followed by stirring at 300 rpm for 15 min. The resulting solution was then centrifuged (Sigma 3-18KS) for 2 min at 1500×g. Clear solvent was collected out followed by re-extracting solvent from aqueous phase by the same method as mentioned above. The solvent was then filtered out through anhydrous Na_2SO_4, followed by solvent evaporation by keeping filtered solvent in a beaker in fume hood at overnight. Beaker was then heated at 105°C for 30 min to remove any residual solvent in oil. The total oil content was then determined gravimetrically.

Encapsulation efficiency:

The encapsulation efficiency was determined by following formula

$$EF = \frac{\text{Total oil} - \text{Surface oil}}{\text{Total oil}} \times 100 \qquad (23.2)$$

Microscopic views:

One gram of sample was dispersed in 20 mL of solution. Both the samples were prepared separately. Each dispersion sample was then observed in Nikon eclipse ci microscope.

Fourier transform infrared spectroscopy (FTIR) of encapsulated samples:

FTIR of encapsulated sample was carried out by BRUKER, ALPHA, ECO-ATR.

Thermogravimetric analysis (TGA) of encapsulated sample:

TGA of encapsulated sample was carried out by NETZSCH, Libra, TG 209 F1.

23.4　RESULT AND DISCUSSION

Encapsulated fish oil has generated a lot of interest among food scientist and food processing industries due to its capability of delivering EFAs in a wide range of food products. Therefore, the technology enables development of a whole range of nutritionally rich beverages, soups, bakery products, ready-to-eat (RTE) snacks, and many more innovative products. Some reports are already available on encapsulation of fish oil with various encapsulating materials and encapsulation process. However, each encapsulation matrix and process results in encapsulated oil with different properties. Some have better powder properties like flow-ability, dispersibility where as some are functionally better, but there is no universally accepted encapsulation technology for all kind of food product. Hence, exploring new encapsulation technology is important for suggesting better process for delivery of fish oil into a challenging food product. In this study, two different matrix materials, pectin and combination of gelatin and pectin, were used for encapsulation of fish oil. In addition, emulsion preparation was done by microfluidizer which provides stable emulsion with mono-disperse type of droplet-size distribution.

23.4.1　MOISTURE CONTENT AND WATER ACTIVITY

Moisture content of pectin encapsulate was 6.4% whereas moisture content of gelatin–pectin composite encapsulate was found out to be 5.5725%. Water activity of pectin encapsulate was 0.23 whereas water activity of gelatin–pectin composite encapsulate was found out to be 0.21. Lower water activity indicates stability against microbial growth, enzymatic, and non-enzymatic activities in particles.

23.4.2 COLOR

Color of pectin encapsulate was L*: 88.11, a*: 1.6, b*: 11.05, and ΔE*: 10.04 whereas color of gelatin–pectin composite encapsulate was found out to be L*: 90.165, a*: 0.885, b*: 11.6, ΔE*: 8.99.

23.4.3 DENSITY

Bulk density of pectin encapsulate was 333.33 kg/m^3 whereas bulk density of gelatin–pectin composite encapsulate was found out to be 375 kg/m^3. Tapped density of pectin encapsulate was 600 kg/m^3 whereas tapped density of gelatin–pectin composite encapsulate was found out to be 545.45 kg/m^3. True density of pectin encapsulate was 1420 kg/m^3 whereas true density of gelatin–pectin composite encapsulate was found out to be 1250 kg/m^3.

23.4.4 ENCAPSULATION EFFICIENCY

Surface oil loading for pectin encapsulate and gelatin–pectin composite encapsulate was found to be 0.10081 and 0.0449 g/g of encapsulated powder, respectively. Thus from above result, we can conclude that surface oil loading for gelatin–pectin composite encapsulate was less compared to pectin encapsulate. Total oil content for pectin encapsulate and gelatin–pectin composite encapsulate was found to be 0.17355 and 0.21905 g/g of encapsulated powder, respectively. Thus from above result, we can conclude that total oil content of gelatin–pectin composite encapsulate was more compared to pectin encapsulate. Encapsulation efficiency of pectin encapsulate and gelatin–pectin composite encapsulate was found to be 41.91 and 79.50%, respectively.

23.4.5 DISPERSIBILITY OF PECTIN ENCAPSULATES AND GELATIN–PECTIN COMPOSITE ENCAPSULATES

Dispersion of encapsulated powder in distilled water showed apparent stability of the rehydrated suspension and there was no significant phase separation even after 24 h storage of the dispersion at room temperature. The fact was further confirmed by the microscopic figure where no significant change in dispersity of the encapsulated particles could be seen (Fig. 23.1).

TABLE 23.1 An Experimental Verification of the Most Promising Reviewed for EFA.

Oil	Wall material	Method	Oil encapsulated/EC	Sources
Fish oil (50% DHA)	Whey protein isolate (WPI)	Spray drying	OE: 33–90% on dry basis	[9]
Fish oil (EPA: DHA, 33:22)	SSPS, maltodextrin, hydroxypropyl betacy-clodextrin, and octenyl succinic anhydride	Spray granulation, spray drying, and freeze drying	EC:SG (96%) > SD (83.62) > FD (<50%)	[10]
Fish oil	N-lauroyl chitosan	Freeze drying	EC: 62.6%	[11]
Fish oil	Sodium caseinate and high amylose resistant starch	Freeze drying	50% oil: 25% protein: 25% starch	[12]
Fish oil	Sodium caseinate and high amylose resistant starch	Freeze drying	50% oil: 25% protein: 25% starch	[13]
Tuna oil	Chitosan, maltodextrin, and whey protein isolate	Freeze drying	EC: 79.3–83.54%	[14]
Fish oil	Caseinate, whey protein-based MRP, and non-MRP-based encapsulant materials	Spray drying	EC: >97%	[15]
Fish oil, and extra virgin olive oil	Sugar beet pectin, and dried glucose syrup	Spray drying	EC: >90%	[16]
Cod liver oil, and omega-3 rich oil	Fish gelatin, chitosan, microbial transglutaminase, and maltodextrin	Spray drying		[17]

EC: Encapsulating efficiency, SG: Spray granulation, SD: Spray drying, FD: Freeze drying, and OE: Oil encapsulated.

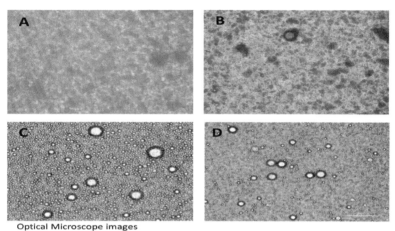

Optical Microscope images
A) Pectin based encapsulates dissolved in distilled water
B) Pectin based encapsulates dissolved in distilled water after 24 h
C) Gelatin based encapsulates dissolved in distilled water
D) Gelatin based encapsulates dissolved in distilled water after 24 h

FIGURE 23.1   Microscopic images of encapsulates dissolved in water.

23.4.6 FTIR SPECTRA

Above FTIR spectra (Fig. 23.2) can be used as a signature spectrum for pectin encapsulates and gelatin–pectin composite encapsulates. This gives the characteristic spectrum of both encapsulates. Also, the peaks of pectin encapsulates between 500 and 1250 wavenumber diminishes in gelatin–pectin composite which indicates that the corresponding functional groups are probably involved in the gelatin–pectin interaction.

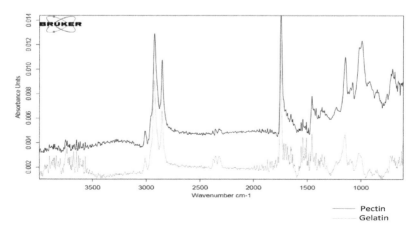

FIGURE 23.2   FTIR spectra of pectin- and gelatin-based encapsulates.

23.4.7 THERMAL CHARACTERISTICS OF FISH OIL ENCAPSULATE

TGA data (Fig. 23.3) of encapsulates showed, in presence of maltodextrin, that the stability of the pectin increases. The degradation slope of pectin encapsulate is lesser than pectin itself and whereas degradation slope of gelatin–pectin composite encapsulate is still lesser than pectin encapsulates. Thus, gelatin and pectin composite encapsulates have grater thermal stability than pectin encapsulates. Therefore, it can be assumed that in high temperature, processing the pectin–gelatin composite encapsulate will have greater promise.

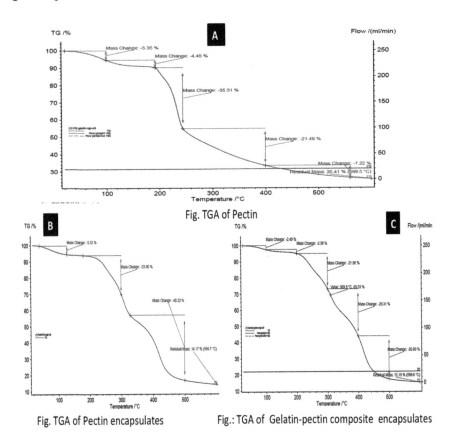

Fig. TGA of Pectin

Fig. TGA of Pectin encapsulates

Fig.: TGA of Gelatin-pectin composite encapsulates

FIGURE 23.3   TGA of encapsulates: (A) represents TGA of pectin, (B) represents TGA of pectin encapsulates, and (C) represents TGA of gelatin–pectin encapsulates.

23.5 CONCLUSION

Microfluidizer can be used efficiently to create microdispersion of fish oil with considerably uniform particle size. Followed by which a freeze drying step can convert encapsulated fish oil into dry powder (~6% moisture content) of longer shelf life. Using pectin or pectin–gelatin composite, matrix fish oil can be delivered as a stable dispersion. Therefore, this work suggests a strategy to encapsulate fish oils to deliver EFAs into challenging food matrices like beverage with low pH, instant soup mixes with better dispersibility, bakery product exposed to harsh baking temperature >170°C, etc. Further application studies of these encapsulates are in progress which will be communicated separately.

KEYWORDS

- essential fatty acids
- freeze drying
- homogenization
- microencapsulation
- microfluidizer
- pectin encapsulate
- gelatin pectin conjugate

REFERENCES

1. McClements, D. J. *Food Emulsions: Principles, Practices and Techniques;* 2nd ed. CRC Press: USA, 2005.
2. Strawbridge, K. R. Measurement of the Particle Size Distributions in Milk Homogenized by a Microfluidizer-Estimation of Populations of Particles with Radii Less than 100 nm. *J. Colloid Interface Sci.* **1995,** *171,* 392.
3. Zuidam, N. N. *Encapsulation Technologies for Active Food Ingredients and Food Processing;* Springer: New York, 2010.
4. Loncin, M. M. *Food Engineering: Principles and Selected Applications;* Academic Press: New York, 1979.
5. Gopal, E. Principles of Emulsion Formation. In *Emulsion Science;* Sherman, P., Ed.; Academic Press: London, UK, 1968; Chap. 1.
6. Schubert, H. Advances in the Mechanical Production of Food Emulsion. In *Engineering and Food;* Jowitt, R., Ed.; Sheffield Academic Press: Sheffield, UK, 1997; p AA82.

7. Stang, M. S. Emulsification in High-Pressure Homogenizers. *Eng. Life Sci.* **2001,** *1,* 151.

8. Karaca, A. C.; Michael, N.; Nicholas, H. L. Microcapsule Production Employing Chickpea or Lentil Protein Isolates and Maltodextrin: Physicochemical Properties and Oxidative Protection of Encapsulated Flaxseed Oil. *Food Chem.* **2013,** *139* (1–4), 448–457. http://dx.doi.org/10.1016/j.foodchem.2013.01.040.

9. Wang, Y.; Wenjie, L.; Xiao Dong, C.; Cordelia, S. Micro-Encapsulation and Stabilization of DHA Containing Fish Oil in Protein-Based Emulsion through Mono-Disperse Droplet Spray Dryer. *J. Food Eng.* **2016,** *175,* 74–84. http://www.sciencedirect.com/science/article/pii/S0260877415300753 (February 2, 2016).

10. Sri Haryani, A.; Benno, K. The Influence of Drying Methods on the Stabilization of Fish Oil Microcapsules: Comparison of Spray Granulation, Spray Drying, and Freeze Drying. *J. Food Eng.* **2011,** *105* (2), 367–378. http://dx.doi.org/10.1016/j.jfoodeng.2011.02.047.

11. Sudipta, C.; Zaher, M. A. J. Impact of Encapsulation on the Physicochemical Properties and Gastrointestinal Stability of Fish Oil. *LWT - Food Sci. Technol.* **2016,** *65,* 206–213. http://dx.doi.org/10.1016/j.lwt.2015.08.010.

12. Cheryl, C.; Luz, S.; Mary Ann, A. In Vitro Lipolysis of Fish Oil Microcapsules Containing Protein and Resistant Starch. *Food Chem.* **2011,** *124* (4), 1480–1489. http://dx.doi.org/10.1016/j.foodchem.2010.07.115.

13. Cheryl, C.; Luz, S.; Mary Ann, A. Effects of Modification of Encapsulant Materials on the Susceptibility of Fish Oil Microcapsules to Lipolysis. *Food Biophys.* **2008,** *3* (2), 140–145.

14. Klaypradit, W.; Yao, W. H. Fish Oil Encapsulation with Chitosan Using Ultrasonic Atomizer. *LWT - Food Sci. Technol.* **2008,** *41* (6), 1133–1139.

15. Kosaraju, S. L.; Rangika, W.; Mary Ann, A. In-Vitro Evaluation of Hydrocolloid-Based Encapsulated Fish Oil. *Food Hydrocol.* **2009,** *23* (5), 1413–1419.

16. Polavarapu, A. S.; Christine, M. O.; Said, A.; Mary Ann, A. Physicochemical Characterisation and Oxidative Stability of Fish Oil and Fish Oil-Extra Virgin Olive Oil Microencapsulated by Sugar Beet Pectin. *Food Chem.* **2011,** *127* (4), 1694–1705. http://dx.doi.org/10.1016/j.foodchem.2011.02.044.

17. Pourashouri, P., et al. Impact of Wall Materials on Physicochemical Properties of Microencapsulated Fish Oil by Spray Drying. *Food Bioprocess Technol.* **2014,** *7* (8), 2354–2365.

INDEX

Printed and bound by CPI Group (UK) Ltd, Croydon, CR0 4YY

23/10/2024

01777701-0007